Memoirs of the American Mathematical Society
Number 225

Joseph A. Wolf

Classification and Fourier inversion for parabolic subgroups with square integrable nilradical

Published by the
AMERICAN MATHEMATICAL SOCIETY
Providence, Rhode Island, USA

November 1979 · Volume 22 · Number 225 (end of volume)

Abstract

In recent years a general theory has been developed for inverting Fourier transforms on non-unimodular locally compact groups. The few known explicit examples have been solvable or have fit into the framework: parabolic subgroup of semisimple Lie group, in which the nilradical has square integrable representations. That class of parabolic subgroups is interesting in its own right; it occurs in many geometric situations, and it has a large overlap with the class of maximal parabolic subgroups.

Here we classify the parabolic subgroups of real and complex semisimple Lie groups, in which the nilradical has square integrable representations. In a few cases -- corresponding to hermitian symmetric spaces of non-tube type -- there is no semi-invariant polynomial on the nilradical. In all other cases we find semi-invariants in the universal enveloping algebra of the nilradical and use them to write out explicit Fourier Inversion formulae.

AMS(MOS) subject classification numbers (1970): Primary 22E30, 22E45. Secondary 43A85, 20G20, 17B35.

Key words and phrases: parabolic subgroup, square integrable representations, semi-invariants, non-unimodular, Fourier inversion formula.

Library of Congress Cataloging in Publication Data

Wolf, Joseph Albert, 1936-
Classification and Fourier inversion for parabolic subgroups with square integrable nilradical.
(Memoirs of the American Mathematical Society ; no. 225)
"Volume 22."
Bibliography: p.
1. Lie groups. 2. Representations of groups. 3. Fourier transformations. 4. Universal enveloping algebras. I. Title. II. Series: American Mathematical Society. Memoirs ; no. 225.
QA3.A57 no. 225 [QA387] 510'.8s [512'.55]
ISBN 0-8218-2225-X 79-21155

CONTENTS

CLASSIFICATION AND FOURIER INVERSION FOR PARABOLIC SUBGROUPS WITH SQUARE INTEGRABLE NILRADICAL

§1. Introduction

In the last ten years, a general theory has been developed for inverting the Fourier transform on a non-unimodular locally compact group. See Tatsuuma [22], Kleppner-Lipsman ([10], [11]) and Duflo-Moore [5]. Pukánszky [19], Moore [17] and Charbonnel [3] have gone into considerable detail for solvable Lie groups.

There are not many explicit examples. The only ones seem to be Kohari's early treatment [12] of the $ax + b$ group, a general understanding ([17], [19], [22]) of the Heisenberg group with scale, Keene's study [8] of Iwasawa subgroups of real rank one Lie groups, and the papers of Keene-Lipsman-Wolf [9] and Lipsman-Wolf [14] on maximal parabolic subgroups of classical groups. Those examples fit into a single framework: parabolic subgroups of semisimple Lie groups, in which the nilradical has square integrable representations [18]. Here square integrability of the nilradical is a strong simplifying assumption.

Received by the Editors: November, 1978

Research partially supported by NSF Grant MCS 76-01692

The Fourier Inversion formula on a type I locally compact group P is of the form

$$(1.1) \qquad f(1_P) = \int_{\hat{P}} \text{trace}\ \pi(DF)d\mu(\pi)$$

where D is an invertible positive self-adjoint operator on $L^2(P)$, semi-invariant of weight given by the modular function δ_P, and μ is a regular Borel measure on the unitary dual $\hat{P}$. When P is a parabolic subgroup with Langlands decomposition MAN in a semisimple or reductive Lie group G, when N has a square integrable representations, and when there is an M-invariant polynomial on the center Z of N, this simplifies to a formula

$$(1.2) \qquad F(1_P) = \sum_{1 \leqslant i \leqslant r} \int_{(M_iA_1)^\wedge} \text{trace}\ (\pi_{i,\nu}(DF))d\mu_i(\nu)$$

in which we explicitly identify all the ingredients. Here D is defined through partial Fourier transform on the Z-variables by the invariant polynomial, say ψ. There are only a finite number r of M-orbits on the hypersurface $|\psi| = 1$, and the isotropy subgroups M_i of M on those orbits are reductive. A_1 is the A-centralizer of Z, so M_iA_1 is reductive with Plancherel measure μ_i known from the Harish-Chandra theory. The class $[\pi_{i,\nu}] \in \hat{P}$ is defined by a class $[\pi_i] \in \hat{N}$ corresponding to the i-th M-orbit using the Kirillov theory, a class $[\nu] \in (M_iA_1)^\wedge$ on the Mackey little-group, and unitary induction from NM_iA_1 to P. This all is done quite explicitly, along the lines of [14], based on explicit knowledge of the structure of P. Thus, in the classification we are careful to make that structure explicit.

In §§2 and 3 we recall the structure theory [18] for square integrable representations of nilpotent Lie groups, and reformulate it in terms of root structure for nilradicals of parabolic subgroups. We also prove some results needed for the classification of parabolic subgroups $P = MAN$ of simple Lie groups G, where N has square integrable representations. The first of those results, Theorem 2.1, says that N has square integrable representations just when $N_{\mathbb{C}}$ has square integrable representations. That reduces the classification to the case where G is $\mathbb{R}$-split (= real normal form = Chevalley group). This reduction is explicitly translated into root structure at the end of §3. The other results, Lemmas 3.8 and 3.9, are techniques for identifying the non-square-integrable cases.

§§4 and 5 give the classification when G is a real or complex simple classical Lie group. We first run through the $\mathbb{R}$-split cases in §4, using Lemmas 3.8 and 3.9 to limit the possibilities and using facts about maximal parabolic subgroups [28] to exhibit square integrable representations in the remaining cases. Then in §5 we use Theorem 2.1 to drop the "$\mathbb{R}$-split" condition and we use [28] in order to be explicit about the structure of the parabolics P.

§§6 and 7 give the classification when G is a real or complex exceptional simple Lie group. The matrix methods at the foundation of [28] are no longer available. Instead, we have to deal with lists of roots to exhibit square integrable representations in the cases not eliminated by Lemmas 3.8 and 3.9, and the book-keeping can be tedious. Matrix methods are not readily available to elicit the explicit structure of P here, and we fall back on a miscellany of facts about finite dimensional representations.

In §8 we note that, from the classification, the existence of square integrable representations of N forces N to be abelian or 2-step nilpotent. Had we known that à priori it would have simplified the classification. We also note an interesting pattern concerning restricted root systems of type F_4, which adds to the list of their intriguing properties.

§9 is a reformulation of the method that Lipsman and I used in [14] to write out Fourier Inversion formulae for certain maximal parabolic subgroups of classical groups. As reformulated, it applies to all the parabolics P = MAN where N has square integrable representations and there is an M-invariant polynomial on the center Z of N, giving explicit inversion formulae in the format (1.2). Then, in Theorem 9.15, we settle the question of existence of the M-invariant polynomial.

In §§10-17 we systematically carry out the program of §9, making (1.2) explicit for each of the parabolics in question. When N is not commutative the Pfaffian polynomial (as in [18]) on the real linear dual $\mathfrak{z}^*$ of the Lie algebra of Z is an M-invariant polynomial, but generally we find a more convenient one. It gives a simpler expression for the pseudo-differential operator D and better information as to whether D is differential. When N is commutative, it is a Jordan algebra and MA is its structure group; see §16.2 for the exceptional case (I didn't see any point to introducing the Jordan algebra machinery for the classical case). This allows us to write down a convenient M-invariant determinant function on $\mathfrak{z}^*$. Once the M-invariant polynomial, say ψ, and the operator, D, are given, we look to the M-orbit structure of $S = \{\lambda \in \mathfrak{z}^* : |\psi(\lambda)| = 1\}$. In most cases that is just some linear algebra, more often straightforward than tricky, but a few cases in §16 require a close look (§16.2) at questions of isotopy of exceptional simple Jordan algebras. When the M-orbit structure of S is worked out we come to a Mackey obstruction problem. Usually it can be settled by methods I developed in [27] and [28], and the remaining cases are amenable to a method worked out in §13.2. That done, the program of §9 immediately gives the Fourier Inversion formula.

A certain air of repetition in §§10-17 seems unavoidable because the details vary so much from case to case.

An error in [28] regarding $O(n,n)$ is corrected in §4.4.1.

I wish to thank Adam Korányi for discussions that resulted in Lemma 9.14 and to thank Daniel Drucker for a helpful correspondence on exceptional real Jordon algebras.

§2. Square Integrability for Nilpotent Groups

We recall some facts on square integrable representations of nilpotent groups, and draw a consequence needed for the classification in §§4 through 7.

Let N be a unimodular locally compact group of type I and let Z denote the center of N. Then the space $\hat{N}$ of equivalence classes of irreducible unitary representations of N is partitioned by $\hat{Z}$, the space of unitary characters on Z,

$$\hat{N} = \bigcup_{\zeta\in\hat{Z}} \hat{N}_\zeta \quad \text{where} \quad \hat{N}_\zeta = \{[\pi] \in \hat{N} : \ \pi \text{ has central character } \zeta\} .$$

Given $\zeta \in \hat{Z}$, let $\ell_\zeta = \mathrm{Ind}_Z^N(\zeta)$, left regular representation of N on the Hilbert space

$$L_2(N/Z,\zeta) = \{f: N \to \mathbb{C}: f(nz) = \zeta(z)^{-1}f(n), \int_{N/Z} |f(n)|^2 d(nZ) < \infty\} .$$

Then N has left regular representation

$$\ell_N = \int_{\hat{Z}} \ell_\zeta d\zeta \quad \text{on} \quad L_2(N) = \int_{\hat{Z}} L_2(N/Z,\zeta)d\zeta \quad .$$

Here $d\zeta$ is Haar measure on $\hat{Z}$, and the Segal-Mautner-Plancherel measure on $\hat{N}$ is of the form $d\mu = d\zeta d\mu_\zeta$ where $L_2(N/Z,\zeta) = \int_{\hat{N}_\zeta} H_\pi \otimes \overline{H_\pi}\, d\mu_\zeta(\pi)$.

A class $[\pi] \in \hat{N}$ is ζ-discrete if π is equivalent to a subrepresentation of ℓ_ζ. The ζ-discrete classes form the ζ-discrete series $\hat{N}_{\zeta\text{-disc}}$, a subset of $\hat{N}_\zeta$. The relative discrete series of N is $\hat{N}_{\text{disc}} = \cup_{\zeta \in \hat{Z}} \hat{N}_{\zeta\text{-disc}}$.

Here are the basic general facts on relative discrete series. Let $[\pi] \in \hat{N}_\zeta$. Given ξ, η in the representation space H_π, denote the coefficient by $\varphi_{\xi,\eta}(n) = \langle \xi, \pi(n)\eta \rangle$. Note that

$$|\varphi_{\xi,\eta}| \in L_2(N/Z) \Longleftrightarrow \varphi_{\xi,\eta} \in L_2(N/Z,\zeta).$$

The following are equivalent:

(i) there exists $0 \neq \xi \in H_\pi$ with $\varphi_{\xi,\xi} \in L_2(N/Z,\zeta)$;

(ii) if ξ, $\eta \in H_\pi$ then $\varphi_{\xi,\eta} \in L_2(N/Z,\zeta)$;

(iii) $[\pi] \in \hat{N}_{\zeta\text{-disc}}$.

Because of this, a class $[\pi] \in \hat{N}_{\text{disc}}$ is called square integrable. Further, if $[\pi], [\pi'] \in \hat{N}_{\zeta\text{-disc}}$ one has positive real numbers $\deg(\pi)$ and $\deg(\pi')$, their formal degrees, such that

$$\langle \varphi_{\xi,\eta}, \varphi_{\xi',\eta'} \rangle = \begin{cases} \deg(\pi)^{-1} \langle \xi, \xi' \rangle \overline{\langle \eta, \eta' \rangle} & \text{if} \quad \pi = \pi' \\ \\ 0 \quad \text{if} \quad [\pi] \neq [\pi'] \end{cases}$$

whenever ξ, $\eta \in H_\pi$ and $\xi', \eta' \in H_{\pi'}$. Here $\mu_\zeta([\pi]) = \deg(\pi)$ as in the Peter-Weyl Theorem. See [26, §2] for details.

Now suppose that N is a connected, simply connected nilpotent Lie group. Moore and I [18] combined the general theory of relative discrete series with the Kirillov theory. Write $\mathfrak{n}$ and $\mathfrak{z}$ for the Lie algebras of N and Z, $\mathfrak{n}^*$ and $\mathfrak{z}^*$ for their real linear dual spaces, and $[\pi_f]$ for the class in $\hat{N}$ associated to the co-adjoint orbit $Ad^*(N)\cdot f \subset \mathfrak{n}^*$. Notice

$$[\pi_f] \in \hat{N}_\zeta \quad \text{where} \quad \zeta(z) = e^{if(\log z)}.$$

The following are equivalent:

(i) $[\pi_f] \in \hat{N}_{disc}$

(ii) $\dim_{\mathbb{R}} Ad^*(N)\cdot f = \dim_{\mathbb{R}}(\mathfrak{n}/\mathfrak{z})$

(iii) $Ad^*(N)\cdot f = \{f' \in \mathfrak{n}^* : f'|_{\mathfrak{z}} = f|_{\mathfrak{z}}\}$

(iv) $b_f(x,y) = f[x,y]$, bilinear form on $\mathfrak{n}$, has kernel $\mathfrak{z}$.

Under those circumstances, b_f induces a nondegenerate antisymmetric bilinear form on $\mathfrak{n}/\mathfrak{z}$ which we also denote by b_f, and so $\mathfrak{n}/\mathfrak{z}$ must have even dimension. We now detect $\hat{N}_{disc}$ as follows. If $\dim_{\mathbb{R}}\mathfrak{n}/\mathfrak{z}$ is odd then $\hat{N}_{disc}$ is empty, so suppose that $\mathfrak{n}/\mathfrak{z}$ has even dimension $2m$. Fix a volume element ω on $\mathfrak{n}/\mathfrak{z}$. If $\lambda \in \mathfrak{z}^*$ extend it to $f = f_\lambda \in \mathfrak{n}^*$. Then b_f is a 2-form on $\mathfrak{n}/\mathfrak{z}$, so its m-th exterior power is a multiple $Pf(b_f)\omega$ of ω. That is the Pfaffian, and it is a well-defined polynomial function $\mathcal{P}(\lambda)$ on $\mathfrak{z}^*$. Now $\mathcal{P}(\lambda) \neq 0$ if and only if $\pi_f \in \hat{N}_{disc}$. So, either $\mathcal{P}$ is identically zero and $\hat{N}_{disc}$ is empty, or $\mathcal{P}$ is not identically zero and $\hat{N}\setminus\hat{N}_{disc}$ has Plancherel measure zero. In the latter case we say that "N has square integrable representations." When N has square integrable representations, $|\mathcal{P}(\lambda)|$ is the formal degree up to a constant depending on $\dim_{\mathbb{R}}\mathfrak{n}/\mathfrak{z}$, normalizations of Haar measures and choice of ω. Then the Plancherel formula has a particularly simple form,

$$\varphi(1_N) = c\int_{z^*} \operatorname{trace} \pi_\lambda(\varphi)|P(\lambda)|d\lambda ,$$

where $d\lambda$ is Lebesgue measure on $\mathfrak{z}^*$.

We will need the following result for the classification in §§4 through 7.

2.1. Theorem. *Let N be a connected simply connected nilpotent Lie group. Let $\tilde{N}$ denote the complexification $N_{\mathbb{C}}$ viewed as a real Lie group. Then N has square integrable representations if and only if $\tilde{N}$ has square integrable representations.*

Proof. Let $\mathfrak{x}$ be a complement to $\mathfrak{z}$ in $\mathfrak{n}$. Then $\tilde{N}$ has Lie algebra $\tilde{\mathfrak{n}} = \tilde{\mathfrak{x}} + \tilde{\mathfrak{z}}$ where $\tilde{}$ denotes the underlying real structure of the complexification and $\tilde{\mathfrak{z}}$ is the center of $\tilde{\mathfrak{n}}$.

Suppose that N has square integrable representations. Choose $\lambda \in \mathfrak{z}^*$ with $b_\lambda(x,y) = \lambda[x,y]$ nonsingular on $\mathfrak{n}/\mathfrak{z}$, where λ is extended to $\mathfrak{n}$ by $\lambda(\mathfrak{x}) = 0$. Define $\tilde{\lambda} \in \tilde{\mathfrak{z}}^*$ by $\tilde{\lambda}(z_1 + iz_2) = \lambda(z_1)$ where $z_1,z_2 \in \mathfrak{z}$, and extend it to $\tilde{\mathfrak{n}}$ by zero on $\tilde{\mathfrak{x}}$. The corresponding bilinear form on $\tilde{\mathfrak{n}}/\tilde{\mathfrak{z}}$ is given by $\tilde{b}_\lambda(x_1 + ix_2, y_1 + iy_2) = \lambda[x_1,y_1] - \lambda[x_2,y_2]$. If $\{x_1,\ldots,x_k\}$ is a basis of $\mathfrak{x}$, and $\{x_1,\ldots,x_{2k}\}$ is the corresponding basis of $\tilde{\mathfrak{x}}$ given by $x_{k+j} = ix_j$, then $\tilde{b}_\lambda$ has matrix $\begin{pmatrix} A & 0 \\ 0 & -A \end{pmatrix}$ where A is the matrix of b_λ. As A is nonsingular, so is $\tilde{b}_\lambda$. Thus $\tilde{N}$ has square integrable representations.

Suppose that $\tilde{N}$ has square integrable representations. Let $\tilde{\lambda} \in \tilde{\mathfrak{z}}^*$ such that the associated bilinear form $\tilde{b}_\lambda$ is nonsingular on $\tilde{\mathfrak{n}}/\tilde{\mathfrak{z}}$. In bases as above, $\tilde{b}_\lambda$ has matrix $\begin{pmatrix} A & B \\ B & -A \end{pmatrix}$ where $a_{rs} = \tilde{\lambda}[x_r,x_s]$ and

$b_{rs} = \tilde{\lambda}[x_r, x_{k+s}]$. Define $\lambda_1, \lambda_2 \in \mathfrak{z}^*$ by

$$\lambda_1(z) = \tilde{\lambda}(z) \text{ and } \lambda_2(z) = \tilde{\lambda}(iz) \text{ for } z \in \mathfrak{z}.$$

Then

$$\lambda_1[x_r, x_s] = \tilde{\lambda}[x_r, x_s] = a_{rs} \quad \text{and}$$

$$\lambda_2[x_r, x_s] = \tilde{\lambda}(i[x_r, x_s]) = \tilde{\lambda}[x_r, x_{k+s}] = b_{rs}$$

so b_λ has matrix $A + tB$ where $\lambda = \lambda_1 + t\lambda_2 \in \mathfrak{z}^*$. Now we need only prove the elementary

2.2. Lemma. *If A and B are $k \times k$ real matrices with* $\begin{pmatrix} A & B \\ B & -A \end{pmatrix}$ nonsingular, then $A + tB$ is nonsingular for some real t.

Proof. $\begin{pmatrix} A & B \\ B & -A \end{pmatrix}$ is row-equivalent to $\begin{pmatrix} B & -A \\ A & B \end{pmatrix}$, which thus is nonsingular, and which is the image of $B - iA$ under $\mathbb{C}^{k\times k} \hookrightarrow \mathbb{R}^{2k\times 2k}$. Now $A + iB = i(B-iA)$ is nonsingular, so the real polynomial $p(t) = \det(A + tB)$ is not identically zero on $\mathbb{C}$, hence not identically zero on $\mathbb{R}$.

qed.

2.3. Corollary. *Let N and N' be connected simply connected nilpotent Lie groups with $\mathfrak{n}_{\mathbb{C}} \cong \mathfrak{n}'_{\mathbb{C}}$. Then N has square integrable representations if and only if N' has square integrable representations.*

§3. Square Integrability for Nilradicals

We reformulate the material of §2, for nilradicals of parabolic subgroups, in terms of roots.

Let $\mathfrak{g}$ be a reductive real Lie algebra. Fix a Cartan involution θ of $\mathfrak{g}$, consider the corresponding Cartan decomposition $\mathfrak{g} = \mathfrak{k} + \mathfrak{s}$, and choose a maximal abelian subspace $\mathfrak{a} \subset \mathfrak{s}$. Then we have

$$\mathfrak{g} = (\mathfrak{m} + \mathfrak{a}) + \sum_{\alpha \in \Delta_{\mathfrak{a}}} \mathfrak{g}_\alpha$$

where $\mathfrak{m} \oplus \mathfrak{a}$ is the centralizer of $\mathfrak{a}$ in $\mathfrak{g}$, $\theta(\mathfrak{m}) = \mathfrak{m}$, and the $\mathfrak{g}_\alpha = \{x \in \mathfrak{g}: [a,x] = \alpha(a)x, \text{ all } a \in \mathfrak{a}\} \neq 0$ are the $\mathfrak{a}$-root spaces. $\Delta_{\mathfrak{a}} \subset \mathfrak{a}^* \setminus \{0\}$ is the $\mathfrak{a}$-<u>root system</u> or <u>restricted root system</u> of $\mathfrak{g}$.

Fix a positive $\mathfrak{a}$-root system $\Delta_{\mathfrak{a}}^+$ and let $B = \{\beta_1, \ldots, \beta_\ell\}$ denote the corresponding simple root system. Whenever $\Xi \subset B$ we denote

$$\langle \Xi \rangle = \{\alpha \in \Delta_{\mathfrak{a}} : \alpha \text{ linear combination of roots in } \Xi\}.$$

Now fix a subset $\Phi \subset B$ and write Φ' for its complement $B \setminus \Phi$. We have algebras

$$(3.1) \qquad \begin{cases} \mathfrak{a}_\Phi = \{a \in \mathfrak{a}: \beta(a) = 0 \text{ for all } \beta \in \Phi'\}, \\ \mathfrak{m}_{\Phi'} = \mathfrak{m} + \mathfrak{a}_{\Phi'} + \sum_{\alpha \in \langle \Phi' \rangle} \mathfrak{g}_\alpha, \\ \mathfrak{n}_\Phi = \sum_{\alpha \in \Delta_{\mathfrak{a}}^+ \setminus \langle \Phi' \rangle} \mathfrak{g}_\alpha, \\ \mathfrak{p}_\Phi = \mathfrak{m}_{\Phi'} + \mathfrak{a}_\Phi + \mathfrak{n}_\Phi . \end{cases}$$

The algebras $\mathfrak{p}_\Phi$ are the parabolic subalgebras of $\mathfrak{g}$ containing the minimal parabolic

(3.2) $$\mathfrak{p}_B = \mathfrak{m} + \mathfrak{a} + \mathfrak{n}, \quad \mathfrak{n} = \sum_{\alpha \in \Delta_{\mathfrak{a}}^+} \mathfrak{g}_\alpha .$$

$\mathfrak{m}_{\Phi'} \oplus \mathfrak{a}_\Phi$ is the $\mathfrak{g}$-centralizer of $\mathfrak{a}_\Phi$, $\mathfrak{m}_{\Phi'}$ is a θ-stable reductive subalgebra of $\mathfrak{g}$ with simple restricted root system Φ', and $\mathfrak{n}_\Phi$ is the nilradical of $\mathfrak{p}_\Phi$.

If $\alpha \in \Delta_{\mathfrak{a}}$ with $\alpha|_{\mathfrak{a}_\Phi} \neq 0$ we will call $\alpha|_{\mathfrak{a}_\Phi}$ an $\mathfrak{a}_\Phi$-root of $\mathfrak{g}$. Let Δ_Φ denote the system of $\mathfrak{a}_\Phi$-roots and define

(3.3) $$\Delta_\Phi^+ = \{\alpha|_{\mathfrak{a}_\Phi} : \alpha \in \Delta_{\mathfrak{a}}^+ \text{ and } \alpha|_{\mathfrak{a}_\Phi} \neq 0\}$$

to be the corresponding positive $\mathfrak{a}_\Phi$-root system. Note that $\Delta_{\mathfrak{a}}^+ = \Delta_B^+$. For convenience of notation, we will identify an element $\alpha|_{\mathfrak{a}_\Phi} \in \Delta_\Phi^+$ with

(3.4) $$[\alpha] = \{\alpha' \in \Delta_{\mathfrak{a}}^+ : \alpha'|_{\mathfrak{a}_\Phi} = \alpha|_{\mathfrak{a}_\Phi}\} = \{\alpha' \in \Delta_{\mathfrak{a}}^+ : \alpha - \alpha' \in \langle\Phi'\rangle\} .$$

Then $\mathfrak{n}_\Phi$ is the sum of the positive $\mathfrak{a}_\Phi$-root spaces,

(3.5) $$\mathfrak{n}_\Phi = \sum_{\Delta_\Phi^+} \mathfrak{n}_{[\alpha]} \quad \text{where} \quad \mathfrak{n}_{[\alpha]} = \sum_{\alpha' \in [\alpha]} \mathfrak{g}_{\alpha'} .$$

If $\Phi = \{\beta_j : j \in J\}$ where $J \subset \{1,\dots,\ell\}$, then Δ_Φ^+ consists of all $[\alpha] = \alpha|_{\mathfrak{a}_\Phi}$ such that α is a root $\sum_1^\ell n_i\beta_i$ with $n_j > 0$ for at least one $j \in J$, and then $[\alpha]$ consists of all roots $\sum n_i'\beta_i$ with $n_j' = n_j$ for all $j \in J$.

We now look for conditions on $\mathfrak{n}_\Phi$ necessary in order that the corresponding connected simply connected group N_Φ have square integrable representations. Any decomposition of $\mathfrak{g}$ as direct sum of ideals $\mathfrak{g}^{(i)}$ splits N_Φ as direct product of the groups for the nilradicals $\mathfrak{n}_\Phi \cap \mathfrak{g}^{(i)}$ of the parabolics $\mathfrak{p}_\Phi \cap \mathfrak{g}^{(i)}$, so we may assume $\mathfrak{g}$ simple. The first part of the proof of [14, Lemma 5.8] shows

3.6. Lemma. Suppose that $\mathfrak{g}$ is simple and let ν be the highest $\mathfrak{a}$-root. Then $[\nu] \in \Delta_\Phi^+$ and $\mathfrak{n}_{[\nu]}$ is the center of $\mathfrak{n}_\Phi$.

Let $\mathfrak{t}$ be a Cartan subalgebra of $\mathfrak{m}$, so $\mathfrak{h} = \mathfrak{t} + \mathfrak{a}$ is a maximally split Cartan subalgebra of $\mathfrak{g}$, and let Δ^+ be a positive $\mathfrak{h}_{\mathbb{C}}$-root system on $\mathfrak{g}_{\mathbb{C}}$ such that $\Delta_{\mathfrak{a}}^+ = \{\gamma|_{\mathfrak{a}} : \gamma \in \Delta^+ \text{ and } \gamma|_{\mathfrak{a}} \neq 0\}$. Denote, for $[\alpha] \in \Delta_\Phi^+$,

$$(3.7) \qquad [[\alpha]] = \{\gamma \in \Delta^+ : \gamma|_{\mathfrak{a}_\Phi} = \alpha|_{\mathfrak{a}_\Phi}\},$$

the $\mathfrak{h}_{\mathbb{C}}$-root analog of (3.4).

Now suppose that N_Φ has square integrable representations, and choose $\lambda \in \mathfrak{z}_\Phi^* = \mathfrak{n}_{[\nu]}^*$ with b_λ nonsingular on $\mathfrak{n}_\Phi/\mathfrak{z}_\Phi \cong \sum_{[\nu]\neq[\alpha]\in\Delta_\Phi^+} \mathfrak{n}_{[\alpha]}$. Then b_λ restricts to a nonsingular pairing of $\mathfrak{n}_{[\alpha]}$ with $\mathfrak{n}_{[\nu]-[\alpha]}$, whenever $[\nu] \neq [\alpha] \in \Delta_\Phi^+$. That gives us

3.8. Lemma. Suppose that $\mathfrak{g}$ is simple, ν is the highest $\mathfrak{a}$-root, and N_Φ has square integrable representations. If $[\nu] \neq [\alpha] \in \Delta_\Phi^+$ then one of the following two alternatives holds.

(i) $2\alpha|_{\mathfrak{a}_\Phi} = \nu|_{\mathfrak{a}_\Phi}$, and $[[\alpha]] = \{\gamma_1, \ldots, \gamma_{2m}\}$, for some integer m, in such a way that $\gamma_j + \gamma_{m+j} \in [[\nu]]$ for $1 \leqslant j \leqslant m$.

(ii) $2\alpha|_{\mathfrak{a}_\Phi} \neq \nu|_{\mathfrak{a}_\Phi}$, *there exists* $[\beta] \in \Delta_\Phi^+$ *with* $[\alpha] + [\beta] = [\nu]$, *and there are enumerations* $[[\alpha]] = \{\gamma_1, \ldots, \gamma_m\}$ *and* $[[\beta]] = \{\delta_1, \ldots, \delta_m\}$ *such that each* $\gamma_j + \delta_j \in [[\nu]]$.

In particular, if $\varphi \in \Phi$ then either $[\varphi] = [\nu]$ and N_Φ is abelian, or $[\nu] - [\varphi] \in \Delta_\Phi^+$. We reformulate that using the extended Dynkin diagram for the simple $\mathfrak{j}_{\mathbb{C}}$-root system of Δ^+. Let μ denote the highest $\mathfrak{j}_{\mathbb{C}}$-root; then $\nu = \mu|_{\mathfrak{a}}$. Every $\beta \in B$ is the $\mathfrak{a}$-restriction either of one or of two simple roots. Now let $\varphi \in \Phi$ with $[\varphi] \neq [\nu]$, and choose a simple $\mathfrak{j}_{\mathbb{C}}$-root ψ with $\psi|_{\mathfrak{a}} = \varphi$. As $[\nu]-[\varphi] \in \Delta_\Phi^+$, there is an $\mathfrak{j}_{\mathbb{C}}$-root δ such that $\delta|_{\mathfrak{a}_\Phi} = (\mu-\gamma)|_{\mathfrak{a}_\Phi}$. Now we may suppose

$$-\delta = \psi + \psi_1 + \ldots + \psi_t + (-\mu)$$

where $\psi_1, \ldots, \psi_t$ are simple $\mathfrak{j}_{\mathbb{C}}$-roots such that each $\psi + \psi_1 + \ldots + \psi_s$, $s \leqslant t$, is a root, and each $\psi_s|_{\mathfrak{a}_\Phi} = 0$. We have just proved.

<u>3.9. Lemma</u>. *Suppose that* $\mathfrak{g}$ *is simple,* μ *is the highest* $\mathfrak{j}_{\mathbb{C}}$*-root, and* N_Φ *has square integrable representations. If* $\varphi \in \Phi$ *and* β *is a simple* $\mathfrak{j}_{\mathbb{C}}$*-root that restricts to* φ*, then there is a path from* β *to* $-\mu$ *in the extended Dynkin diagram, which does not pass through any other simple* $\mathfrak{j}_{\mathbb{C}}$*-root whose* $\mathfrak{a}$*-restriction is in* Φ.

Lemma 3.9 will let us radically limit the possibilities for Φ before applying the more delicate criteria of Lemma 3.8.

The general method of passing between real forms is as follows. Suppose that $\mathfrak{g}$ and $\mathfrak{g}'$ are simple real Lie algebras with $\mathfrak{g}_{\mathbb{C}} = \mathfrak{g}'_{\mathbb{C}}$, and that $\mathfrak{h} = \mathfrak{t} + \mathfrak{a}$ and $\mathfrak{h}' = \mathfrak{t}' + \mathfrak{a}'$ are Cartan subalgebras with $\mathfrak{h}_{\mathbb{C}} = \mathfrak{h}'_{\mathbb{C}}$. We have a simple $\mathfrak{h}_{C}$-root system $\Psi = \{\psi_1,\ldots,\psi_m\}$ of $\mathfrak{g}_{\mathbb{C}}$, a simple $\mathfrak{a}$-root system $B = \{\beta_1,\ldots,\beta_\ell\}$ on $\mathfrak{g}$, and a simple $\mathfrak{a}'$-root system $D = \{\delta_1,\ldots,\delta_k\}$ on $\mathfrak{g}'$, such that

if $\psi \in \Psi$ then either $\psi|_{\mathfrak{a}} = 0$ or $\psi|_{\mathfrak{a}} \in B$,

if $\psi \in \Psi$ then either $\psi|_{\mathfrak{a}'} = 0$ or $\psi|_{\mathfrak{a}'} \in D$.

Let $\Phi \subset B$. Then $(\mathfrak{p}_\Phi)_{\mathbb{C}}$ is the parabolic subalgebra of $\mathfrak{g}_{\mathbb{C}}$ for $\tilde{\Phi} = \{\psi \in \Psi\colon \psi|_{\mathfrak{a}} \in \Phi\}$. So $\mathfrak{g}'$ has a parabolic subalgebra $\mathfrak{p}'_\Gamma$ with complexification $(\mathfrak{p}_\Phi)_{\mathbb{C}}$ just when $\tilde{\Phi} = \{\psi \in \Phi\colon \psi|_{\mathfrak{a}'} \in \Gamma\}$ for some subset $\Gamma \subset D$. One checks this by a glance at the Satake diagrams of $\mathfrak{g}$ and $\mathfrak{g}'$. Here recall that the Satake diagram of $\mathfrak{g}$ is obtained from the Dynkin diagram of $\mathfrak{g}_{\mathbb{C}}$ (arrow convention, so F_4 is o—o⇒o—o) by blackening every vertex that corresponds to a root $\psi \in \Psi$ with $\psi|_{\mathfrak{a}} = 0$, and joining by an arrow any pair ψ, ψ' of roots in Ψ with $\psi|_{\mathfrak{a}} = \psi'|_{\mathfrak{a}} \neq 0$ (e.g. 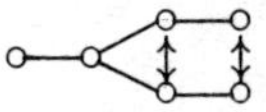). Then Corollary 2.3 tells us either that both of $N_\Phi \subset G$ and $N'_\Gamma \subset G'$ have square integrable representations, or that neither does.

§4. Classification in the Real Split Classical Groups

We apply the results of [28] and of §3 to find all parabolic subgroups of classical simple split real (= normal form) Lie groups, in which the nilradical has square integrable representations. Roughly speaking, results of §3 limit the possibilities and results of [28] exhibit square integrable representations in the remaining cases.

An error in [28] regarding $O(n,n)$ is corrected in (4.4.1) below.

4.1. Split A_{n-1}: $\overset{\psi_1}{\circ}$—$\overset{\psi_2}{\circ}$—...—$\overset{\psi_{n-1}}{\circ}$. We may take $G = SL(n;\mathbb{R})$, acting on $\mathbb{R}^n$ as usual.

4.1.1. $\Phi = \{\psi_j\}$. Then $\Delta_\Phi^+ = \{[\psi_j]\}$ where $[\psi_j]$ consists of all $\psi_i + \psi_{i+1} + \ldots + \psi_k$, $1 \leqslant i \leqslant j \leqslant k < n$ with $i < k$, and of ψ_j. So $\mathfrak{n}_\Phi$ is abelian of dimension $j(n-j)$ and $P_\Phi = \{g \in G: \mathrm{Ad}(g)\mathfrak{p}_\Phi = \mathfrak{p}_\Phi\}$ is the G-stabilizer of a j-dimensional subspace of $\mathbb{R}^n$. In particular N_Φ has square integrable representations.

4.1.2. $\Phi = \{\psi_i, \psi_j\}$, $i < j$. Then $\Delta_\Phi^+ = \{[\psi_i], [\psi_j], [\psi_i] + [\psi_j]\}$, where $[\psi_i]$ consists of the $i(j-i)$ roots $\psi_s + \ldots + \psi_r$, $1 \leqslant s \leqslant i \leqslant r < j$ and $[\psi_j]$ consists of the $(n-j)(j-i)$ roots $\psi_u + \ldots + \psi_v$, $i < u \leqslant j \leqslant v < n$. If $n-j \neq i$, then $\dim \mathfrak{n}_{[\psi_i]} \neq \dim \mathfrak{n}_{[\psi_j]}$, so Lemma 3.8 (ii) shows that N_Φ does not have square integrable representations.

Now suppose $n-j = i$, that is $j = n-i$. For $1 \leqslant s \leqslant i$ denote

$$\gamma_s = \psi_s + \psi_{s+1} + \ldots + \psi_{n-s} \in [\psi_i] + [\psi_{n-i}].$$

Choose nonzero linear functions f_s on $\mathfrak{n}_\Phi$ such that $f_s(\mathfrak{g}_\alpha) = 0$ for $\alpha \neq \gamma_s$ and let $f = f_1 + f_2 + \ldots + f_i$. Partition $[\psi_i]$ into subsets

$$\pi(s) = \{\psi_s + \ldots + \psi_i, \psi_s + \ldots + \psi_{i+1}, \ldots, \psi_s + \ldots + \psi_{n-i-1}\}$$

and partition $[\psi_{n-i}]$ into subsets

$$\pi'(s) = \{\psi_{n-i} + \ldots + \psi_{n-s}, \ldots, \psi_{i+2} + \ldots + \psi_{n-s}, \psi_{i+1} + \ldots + \psi_{n-s}\}.$$

The corresponding parts of $\mathfrak{n}_\Phi$ are the

$$\mathfrak{n}_{\pi(s)} = \sum_{\pi(s)} \mathfrak{g}_\alpha \subset \mathfrak{n}_{[\psi_i]} \quad \text{and} \quad \mathfrak{n}_{\pi'(s)} = \sum_{\pi'(s)} \mathfrak{g}_\alpha \subset \mathfrak{n}_{[\psi_{n-i}]}.$$

The bilinear form

$$b_{f_s} : \mathfrak{n}_{\pi(s)} \times \mathfrak{n}_{\pi'(t)} \to \mathbb{R} \text{ is } 0 \text{ for } s \neq t, \text{ nonsingular for } s = t.$$

Now $b_f : \mathfrak{n}_{[\psi_i]} \times \mathfrak{n}_{[\psi_{n-i}]} \to \mathbb{R}$ is nonsingular. Thus $N_{\{\psi_i, \psi_{n-i}\}}$ has square integrable representations.

The affirmative part of (4.1.2) also follows from [28 , Corollary 4.5] with $F = \mathbb{C}$ and $s = i$ by means of Corollary 2.3; but later we will need to refer to the partitioning method used here.

<u>4.1.3.</u> Φ has 3 or more elements. Then Lemma 3.9 shows that N does not have square integrable representations.

4.2. Split B_n: 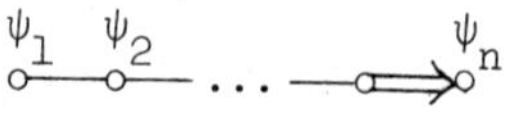

, $n \geqslant 2$. We may take $G = SO(n,n+1)$, all elements of determinant 1 in the orthogonal group of the euclidean vector space $\mathbb{R}^{n,n+1}$ with inner product $\langle x,y\rangle = -\sum_1^n x_jy_j + \sum_{n+1}^{2n+1} x_jy_j$.

4.2.1. $\Phi = \{\psi_j\}$. Then P_Φ is a maximal parabolic subgroup of G and $\mathfrak{m}_\Phi$, is of type $A_{j-1} \times B_{n-j}$. It follows from [28, Theorem 2.10] with $F^{p,q} = \mathbb{R}^{n,n+1}$ that P_Φ is the G-stabilizer of a totally isotropic j-dimensional subspace of $\mathbb{R}^{n,n+1}$ and is a subgroup of index 2 in the group $P_{j;n-j,n+j-1}(\mathbb{R})$ defined in [28, pp. 26-28]. Thus [28, Corollary 4.5] N_Φ has square integrable representations exactly when $j = 1$ or j is even.

4.2.2. $\Phi = \{\psi_i,\psi_j\}$, $i < j$. The extended Dynkin diagram is

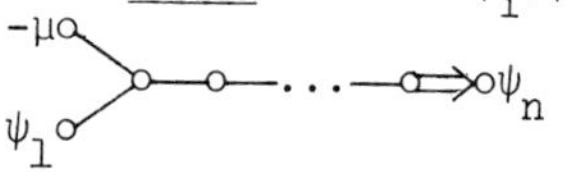

, so Lemma 3.9 says that N_Φ cannot have square integrable representations when $i > 1$ or when $j < 3$.

Now suppose $\Phi = \{\psi_1,\psi_j\}$, $3 \leqslant j \leqslant n$. Then $\Delta_\Phi^+ = \{[\psi_1], [\psi_j], 2[\psi_j], [\psi_1] + [\psi_j], [\psi_1] + 2[\psi_j]\}$. Here $[\psi_j]$ consists of the $(j-1)(n+1-j)$ roots $\psi_r + \ldots + \psi_s$ with $2 \leqslant r \leqslant j \leqslant s \leqslant n$ and the $(j-1)(n-j)$ roots $\psi_r + \ldots + \psi_{s-1} + 2(\psi_s + \ldots + \psi_n)$ with $2 \leqslant r \leqslant j < s \leqslant n$, so $\dim \mathfrak{n}_{[\psi_j]} = (j-1)(2n-2j+1)$. Similarly, $[\psi_1] + [\psi_j]$ consists of the $n-j+1$ roots $\psi_1 + \ldots + \psi_r$ with $j \leqslant r \leqslant n$ and the $n-j$ roots $\psi_1 + \ldots + \psi_{r-1} + 2(\psi_r + \ldots + \psi_n)$ with $j < r \leqslant n$, so $\dim \mathfrak{n}_{[\psi_1]+[\psi_j]} = 2n-2j+1$. If N_Φ is to have square integrable representations, those dimensions must be equal according to Lemma 3.8. so $j = 2$, contradicting $j \geqslant 3$. Thus N_Φ does not have square integrable representations.

4.2.3. Φ has 3 or more elements. Then Lemma 3.9 shows that N_Φ does not have square integrable representations.

4.3. **Split** C_n : 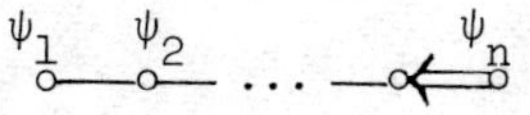 , $n \geq 2$. We may take $G = Sp(n;\mathbb{R})$, real symplectic group, which is the automorphism group of $(\mathbb{R}^{2n}, \{,\})$ where $\{\ ,\ \}$ is a nondegenerate antisymmetric bilinear form.

4.3.1. $\Phi = \{\psi_j\}$. Then P is a maximal parabolic subgroup of G and $\mathfrak{m}_\Phi$, is of type $A_{j-1} \times C_{n-j}$. It follows from [28, Theorem 8.6] with $F = \mathbb{R}$ that P_Φ is the G-stabilizer of a j-dimensional totally $\{\ ,\ \}$-isotropic subspace of $\mathbb{R}^{2n}$ and is isomorphic to the group $P_{j;2(n-j)}(\mathbb{R})$ defined in [28 , pp. 83-84]. Thus [28 , Corollary 9.5] N_Φ has square integrable representations.

4.3.2. Φ has 2 or more elements. Then Lemma 3.9 shows that N_Φ does not have square integrable representations.

4.4. Split D_n : 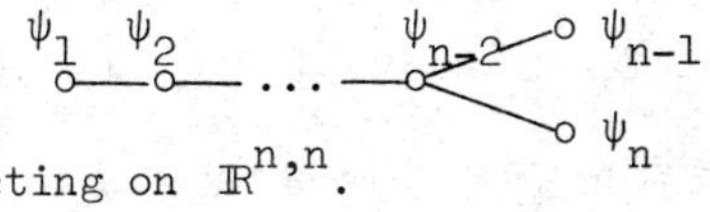 , $n \geq 4$. We may take $G = SO(n,n)$ acting on $\mathbb{R}^{n,n}$.

4.4.1. There is an error in [28] concerning $O(n,n)$ and $SO(n,n)$, and we take this opportunity to correct it. The point is that the $SO(n,n)$-stabilizer P_E of an $(n-1)$-dimensional totally isotropic subspace $E \subset \mathbb{R}^{n,n}$ also stabilizes two n-dimensional totally isotropic subspaces U_1, U_2. Thus P_E is not a maximal parabolic, but rather is the intersection of the two maximal parabolics P_{U_1} and P_{U_2}, corresponding to the cases $\Phi = \{\psi_{n-1}\}$ and $\Phi = \{\psi_n\}$ and thus $O(n,n)$-conjugate but not $SO(n,n)$-conjugate. This is seen as follows. $E^\perp = E \oplus W$ with $W \cong \mathbb{R}^{1,1}$ as in [28, (3.1)] with $s = n-1$ and $F^{p,q} = \mathbb{R}^{n,n}$. The reductive part P_E^r stabilizes W, acting there as $SO(1,1)$ and thus preserving its two isotropic lines W_1, W_2, by [28, Lemmas 3.6 and 3.13]; and the nilradical $\mathfrak{n}_E$ maps W into E by [28, Lemma 3.7]. Now set $U_i = E + W_i$. See below for the connection to Φ.

There also is a trivial error in [28, Corollary 4.5]. In addition to the cases listed, $N_{n;0,0}(\mathbb{R})$, nilradical of the stabilizer of an n-dimensional totally isotropic subspace of $\mathbb{R}^{n,n}$, is abelian and thus has square integrable representations, even when n is odd.

4.4.2. $\Phi = \{\psi_j\}$, $j \neq n-1$. Then P_Φ is a maximal parabolic subgroup of G and $\mathfrak{m}_\Phi$, is of type $A_{j-1} \times D_{n-j}$. It follows from [28, Theorem 2.10] with $F^{p,q} = \mathbb{R}^{n,n}$ that P_Φ is the G-stabilizer of a totally isotropic j-dimensional subspace of $\mathbb{R}^{n,n}$ and is locally isomorphic to the group $P_{j;n-j,n-j}(\mathbb{R})$ defined in [28, pp. 26-28]. Thus [28, Corollary 4.5] N_Φ has square integrable representations precisely in the cases $j = 1$ and $j = n$ where it is abelian, and in the cases where j is even.

4.4.3. $\Phi = \{\psi_{n-1}\}$. Then $P_\Phi \cong P_{\{\psi_n\}}$, so N has square integrable representations.

4.4.4. $\Phi = \{\psi_{n-1}, \psi_n\}$. From the discussion above, $P_\Phi = P_{\{\psi_{n-1}\}} \cap P_{\{\psi_n\}}$, the G-stabilizer of an (n-1)-dimensional totally isotropic subspace of $\mathbb{R}^{n,n}$, locally isomorphic to $P_{n-1;1,1}(\mathbb{R})$. Thus [28, Corollary 4.5] N_Φ has square integrable representations just when n-1 is even, i.e. when n is odd.

4.4.5. $\Phi = \{\psi_i, \psi_j\} \neq \{\psi_{n-1}, \psi_n\}$ with $i < j$. The extended Dynkin diagram is 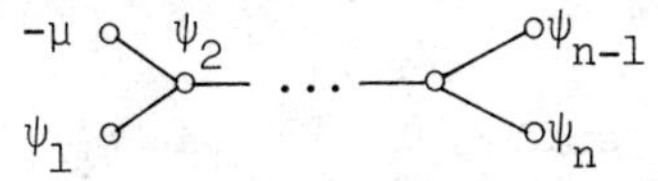 ; so Lemma 3.9 says that N_Φ cannot have square integrable representations except, possibly, when $i = 1$ and $j \geqslant 3$.

First suppose $\Phi = \{\psi_1, \psi_j\}$ where $3 \leqslant j \leqslant n-2$. Then $\Delta_\Phi^+ = \{[\psi_1], [\psi_j], 2[\psi_j], [\psi_1] + [\psi_j], [\psi_1] + 2[\psi_j]\}$. For every root $\psi_1 + \psi_2 + \ldots + \psi_j + \beta \in [\psi_1] + [\psi_j]$ we have $j-1 \geqslant 2$ roots $\psi_r + \ldots + \psi_{j-1} + \psi_j + \beta \in [\psi_j]$ as $2 \leqslant r \leqslant j$. Thus $\dim \mathfrak{n}_{[\psi_j]} > \dim \mathfrak{n}_{[\psi_1]+[\psi_j]}$, and Lemma 3.8 says that N_Φ does not have square integrable representations.

Next suppose $\Phi = \{\psi_1, \psi_n\}$. Then $\Delta_\Phi^+ = \{[\psi_1], [\psi_n], [\psi_1]+[\psi_n]\}$. Here $[\psi_1]$ consists of the n-1 roots $\psi_1 + \psi_2 + \ldots + \psi_r$, $r \leqslant n-1$. Also, $[\psi_n]$ consists of the 1 root ψ_n, the n-3 roots $\psi_r + \ldots + \psi_{n-2} + \psi_n$ with $2 \leqslant r \leqslant n-2$, the n-3 roots $\psi_r + \ldots + \psi_{n-2} + \psi_{n-1} + \psi_n$ with $2 \leqslant r \leqslant n-2$, and the $\frac{1}{2}(n-3)(n-4)$ roots $\psi_r + \ldots + \psi_{s-1} + 2(\psi_s + \ldots + \psi_{n-2}) + \psi_{n-1} + \psi_n$ with $2 \leqslant r < s \leqslant n-2$. Now

$$\dim \mathfrak{n}_{[\psi_1]} = n-1 \quad \text{and} \quad \dim \mathfrak{n}_{[\psi_n]} = \tfrac{1}{2}(n-1)(n-2).$$

If N_Φ is to have square integrable representations, then these dimensions must be equal according to Lemma 3.8, so $n = 4$. But if $n = 4$ then we are in the situation of (4.4.4) with even n. Thus in general $N_{\{\psi_1,\psi_n\}}$ does not have square integrable representations.

Finally suppose $\Phi = \{\psi_1,\psi_{n-1}\}$. Then $P_\Phi \cong P_{\{\psi_1,\psi_n\}}$, so N_Φ does not have square integrable representations.

4.4.6. Φ has 3 or more elements. Lemma 3.9 says that N_Φ cannot have square integrable representations except, perhaps, when $\Phi = \{\psi_1,\psi_{n-1},\psi_n\}$.

Let $\Phi = \{\psi_1,\psi_{n-1},\psi_n\}$. Then Δ_Φ^+ consists of $[\psi_1]$, $[\psi_{n-1}]$, $[\psi_n]$, $[\psi_1] + [\psi_{n-1}]$, $[\psi_1] + [\psi_n]$, $[\psi_{n-1}] + [\psi_n]$, and $[\psi_1] + [\psi_{n-1}] + [\psi_n]$. Here $[\psi_1] + [\psi_{n-1}]$ consists of the single root $\psi_1 + \psi_2 + \dots + \psi_{n-2} + \psi_{n-1}$, while $[\psi_n]$ consists of the $n-3$ roots $\psi_r + \psi_{r+1} + \dots + \psi_{n-2} + \psi_n$ for $2 \leqslant r \leqslant n-2$ and the root ψ_n. If N_Φ is to have square integrable representations then Lemma 3.8 says $\dim \mathfrak{n}_{[\psi_1]+[\psi_{n-1}]} = 1$ and $\dim \mathfrak{n}_{[\psi_n]} = n-2$ must be equal, so $n = 3$, contradicting $n \geqslant 4$. Thus N_Φ does not have square integrable representations.

4.5. Summary-classical split case. We summarize, using results of [28] for the global form of P_Φ and N_Φ, and the method of [29, pp. 282-282b] for the representation of $\mathfrak{m}_{\Phi'}$ on the $\mathfrak{n}_{[\alpha]}$. Here are all the parabolic subgroups of real split simple classical groups, in which the nilradical has square integrable representations.

4.5.1. $G = SL(n;\mathbb{R})$: $\overset{\psi_1}{\circ}\!-\!\overset{\psi_2}{\circ}\!-\cdots-\!\overset{\psi_{n-1}}{\circ}$ with $n \geqslant 2$.

(i) $\Phi = \{\psi_s\}$, $1 \leqslant s \leqslant n-1$. Then P_Φ is the stabilizer of an s-dimensional subspace of $\mathbb{R}^n$, and is given by

$$P_\Phi \cong \left\{\begin{pmatrix}\alpha & x\\ 0 & \beta\end{pmatrix}: \alpha \in GL(s;\mathbb{R}), \beta \in GL(n-s;\mathbb{R}), (\det\alpha)(\det\beta) = 1,\ x \in \mathbb{R}^{s\times(n-s)}\right\}.$$

Here N_Φ consists of the $\begin{pmatrix}I & x\\ 0 & I\end{pmatrix}$, and $M_\Phi A_\Phi = \left\{\begin{pmatrix}\alpha & 0\\ 0 & \beta\end{pmatrix}\right\}$ acts on it through conjugation by $\begin{pmatrix}\alpha & 0\\ 0 & \beta\end{pmatrix}: x \mapsto \alpha x\beta^{-1}$. So $N_\Phi \cong \mathbb{R}^{s\times(n-s)}$, abelian, and $\mathfrak{m}_\Phi$ acts on $\mathfrak{n}_\Phi$ by $\underset{\psi_1}{\overset{1}{\circ}}\!-\cdots-\!\underset{\psi_{s-1}}{\circ} \otimes \underset{\psi_{s+1}}{\circ}\!-\cdots-\!\underset{\psi_{n-1}}{\overset{1}{\circ}}$.

(ii) $\Phi = \{\psi_s, \psi_{n-s}\}$, $1 \leqslant s \leqslant \left[\frac{n-1}{2}\right]$. Then P_Φ is the stabilizer of a flag $\mathbb{R}^s \subset \mathbb{R}^{n-s}$ in $\mathbb{R}^n$, and is given by

$$P_\Phi \cong \left\{\begin{pmatrix}\alpha & x & z\\ 0 & \beta & y\\ 0 & 0 & \gamma\end{pmatrix}: \begin{array}{l}\alpha, \gamma \in GL(s;\mathbb{R}),\ \beta \in GL(n-2s;\mathbb{R}),\\ (\det\alpha)(\det\beta)(\det\gamma) = 1,\\ x \in \mathbb{R}^{s\times(n-2s)},\ y \in \mathbb{R}^{(n-2s)\times s},\ z \in \mathbb{R}^{s\times s}\end{array}\right\}$$

Here

$$N_\Phi = \left\{\begin{pmatrix}I & x & z\\ 0 & I & y\\ 0 & 0 & I\end{pmatrix}\right\} \text{ and } \mathfrak{n}_\Phi = \left\{\begin{pmatrix}0 & x & z\\ 0 & 0 & y\\ 0 & 0 & 0\end{pmatrix}\right\}$$

where (in the algebra) x gives $\mathfrak{n}_{[\psi_j]}$, y gives $\mathfrak{n}_{[\psi_{n-j}]}$, and z gives the center $\mathfrak{n}_{[\psi_j]+[\psi_{n-j}]}$. So N_Φ is a generalized Heisenberg group. $M_\Phi A_\Phi = \left\{\begin{pmatrix}\alpha & & \\ & \beta & \\ & & \gamma\end{pmatrix}\right\}$ acts on it through conjugation by

$$\begin{pmatrix}\alpha & & \\ & \beta & \\ & & \gamma\end{pmatrix}: (z,y,x) \mapsto (\alpha z\gamma^{-1}, \beta y\gamma^{-1}, \alpha x\beta^{-1}).$$

Thus $\mathfrak{m}_{\Phi'}$ acts on $\mathfrak{n}_\Phi$ through $\overset{1}{\underset{\psi_1}{\circ}}—\cdots—\underset{\psi_{s-1}}{\circ} \otimes \underset{\psi_{s+1}}{\circ}—\cdots—\underset{\psi_{n-s-1}}{\circ} \otimes \underset{\psi_{n-s+1}}{\circ}—\cdots—\overset{1}{\underset{\psi_{n-1}}{\circ}}$

on $\mathfrak{n}_{[\psi_j]+[\psi_{n-j}]}$, $\overset{1}{\circ}—\cdots—\circ \otimes \circ—\cdots—\overset{1}{\circ} \otimes \circ—\cdots—\circ$ on $\mathfrak{n}_{[\psi_j]}$,

and $\circ—\cdots—\circ \otimes \overset{1}{\circ}—\cdots—\circ \otimes \circ—\cdots—\overset{1}{\circ}$ on $\mathfrak{n}_{[\psi_{n-j}]}$.

4.5.2. $G = SO(n,n+1)$: $\underset{\psi_1}{\circ}—\underset{\psi_2}{\circ}—\cdots—\underset{\psi_{n-1}}{\circ}\Rightarrow\underset{\psi_n}{\circ}$ with $n \geqslant 2$.

(i) $\Phi = \{\psi_s\}$ with $s = 1$ or $s = 2,4,\dots,2\left[\frac{n}{2}\right]$. Then P_Φ is the stabilizer of an s-dimensional totally isotropic subspace of $\mathbb{R}^{n,n+1}$. It is given by $N_\Phi A_\Phi M_{\Phi'}$ as follows.

Skew $\mathbb{R}^{s\times s} = \{z \in \mathbb{R}^{s\times s}: {}^t z = -z\}$, and $\mathbb{R}^{s\times(t,u)}$ denotes $\mathbb{R}^{s\times t} \oplus \mathbb{R}^{s\times u}$ with the Skew $\mathbb{R}^{s\times s}$-valued bilinear form

$$\mathcal{S}((x_1,x_2),(y_1,y_2)) = \tfrac{1}{2}\{x_1\cdot {}^t y_1 - x_2\cdot {}^t y_2 - y_1\cdot {}^t x_1 + y_2\cdot {}^t x_2\} .$$

Now $N_\Phi \cong \text{Skew}\,\mathbb{R}^{s\times s} + \mathbb{R}^{s\times(n-s,n+1-s)} \cong \mathfrak{n}_\Phi$ with respective product laws

$$(z,x)(z',x') = (z' + z' + \mathcal{S}(x,x'),\ x + x') \quad \text{and}$$

$$[(z,x),(z',x')] = (2\mathcal{S}(x,x'),\ 0)$$

Further,

$$M_{\Phi'}A_\Phi \cong GL(s;\mathbb{R}) \times SO(n-s,n+1-s)$$

and it acts on N_Φ and $\mathfrak{n}_\Phi$ by $(\gamma,g):(z,x) \mapsto (\gamma z\cdot {}^t\gamma, \gamma x\cdot {}^t g)$.

If $s = 1$ then N_Φ is abelian, $N_\Phi \cong \mathbb{R}^{n-1,n} \cong \mathfrak{n}_\Phi = \mathfrak{n}_{[\psi_1]}$,
and $\mathfrak{m}_{\Phi'}$ acts on $\mathfrak{n}_\Phi$ by $\overset{1}{\underset{\psi_2}{\circ}}\!-\cdots-\!\circ\!\Rightarrow\!\underset{\psi_n}{\circ}$.

If $s = 2,4,\ldots,2\left[\frac{n}{2}\right]$ then $\mathfrak{n}_\Phi$ is 2-step nilpotent with center $\mathfrak{n}_{2[\psi_s]} \cong \text{Skew } \mathbb{R}^{s\times s}$, and with $\mathfrak{n}_{[\psi_s]} \cong \mathbb{R}^{s\times(n-s,n+1-s)}$. Then $\mathfrak{m}_{\Phi'}$ acts on $\mathfrak{n}_{[\psi_s]}$ by $\underset{\psi_1}{\circ}\!-\!\overset{1}{\underset{\psi_2}{\circ}}\!-\cdots-\!\underset{\psi_{s-1}}{\circ} \otimes \underset{\psi_{s+1}}{\circ}\!-\cdots-\!\circ\!\Rightarrow\!\underset{\psi_n}{\circ}$ if $s > 2$, $\circ \otimes \circ\!-\cdots-\!\circ\!\Rightarrow\!\circ$ (trivially) if $s = 2$; and $\mathfrak{m}_{\Phi'}$ acts on $\mathfrak{n}_{[\psi_s]}$ by $\overset{1}{\circ}\!-\cdots-\!\circ \otimes \overset{1}{\circ}\!-\cdots-\!\circ\!\Rightarrow\!\circ$.

4.5.3. $G = Sp(n;\mathbb{R})$: $\overset{\psi_2}{\circ}\!-\!\overset{\psi_2}{\circ}\!-\cdots-\!\circ\!\Leftarrow\!\overset{\psi_n}{\circ}$ with $n \geqslant 2$.

(i) $\Phi = \{\psi_s\}$ with $1 \leqslant s \leqslant n$. Then P_Φ is the G-stabilizer of a totally isotropic s-dimensional subpsace of $(\mathbb{R}^{2n}, \{,\})$. It is given by $N_\Phi A_\Phi M_{\Phi'}$ as follows.

$\text{Sym } \mathbb{R}^{s\times s} = \{z \in \mathbb{R}^{s\times s} : z = {}^t z\}$, and we have
$\mathfrak{a}: \mathbb{R}^{s\times 2t} \times \mathbb{R}^{s\times 2t} \to \text{Sym } \mathbb{R}^{s\times s}$ given by

$$\mathfrak{a}((x_1,x_2),(y_1,y_2)) = \tfrac{1}{2}\{x_1\cdot{}^t y_2 - y_2\cdot{}^t x_1 + y_1\cdot{}^t x_2 - x_2\cdot{}^t y_1\} .$$

$N_\Phi \cong \text{Sym } \mathbb{R}^{s\times s} + \mathbb{R}^{s\times 2(n-s)} \cong \mathfrak{n}_\Phi$ with product laws

$$(z,x)(z',x') = (z + z' + \mathfrak{a}(x,x'), x + x') \text{ and}$$
$$[(z,x),(z',x')] = (2\mathfrak{a}(x,x'),0).$$

Further

$$M_{\Phi'}A_\Phi \cong GL(s;\mathbb{R}) \times Sp(n-s;\mathbb{R})$$

and it acts on N_Φ and $\mathfrak{n}_\Phi$ by $(\gamma,g):(z,x) \mapsto (\gamma z \cdot {}^t\gamma, \gamma x \cdot {}^t g)$.

If $s = n$ then $N_\Phi \cong \mathrm{Sym}\, \mathbb{R}^{n\times n}$, abelian, and $\mathfrak{m}_{\Phi'}$ acts on $\mathfrak{n}_\Phi = \mathfrak{n}_{[\psi_s]}$ by $\underset{\psi_1}{\overset{2}{\circ}} - \cdots - \underset{\psi_{n-1}}{\circ}$.

If $1 \leqslant s < n$ then $\mathfrak{n}_\Phi$ is 2-step nilpotent with center $\mathfrak{n}_{2[\psi_s]} \cong \mathrm{Sym}\, \mathbb{R}^{s\times s}$ and with $\mathfrak{n}_{[\psi_s]} \cong \mathbb{R}^{s\times 2(n-s)}$. Then $\mathfrak{m}_{\Phi'}$ acts on $\mathfrak{n}_{2[\psi_s]}$ by $\underset{\psi_1}{\overset{2}{\circ}} - \cdots - \underset{\psi_{s-1}}{\circ} \otimes \underset{\psi_{s+1}}{\circ} - \cdots - \circ \Leftarrow \underset{\psi_n}{\circ}$ for $s > 1$ and trivially for $s = 1$; and $\mathfrak{m}_{\Phi'}$ acts on $\mathfrak{n}_{[\psi_s]}$ by $\overset{1}{\circ} - \cdots - \circ \otimes \overset{1}{\circ} - \cdots - \circ \Leftarrow \circ$ for $s > 1$ and $\overset{1}{\circ} - \cdots - \circ \Leftarrow \circ$ for $s = 1$.

4.5.4. $G = SO(n,n)$: $\underset{\psi_1}{\circ} - \underset{\psi_2}{\circ} - \cdots - \circ \lessdot \begin{smallmatrix} \circ\,\psi_{n-1} \\ \circ\,\psi_n \end{smallmatrix}$ with $n \geqslant 4$.

(i) $\Phi = \{\psi_s\}$ with $s = 1$ or $s = 2,4,\ldots,2\left[\frac{n-2}{2}\right]$. Then P_Φ is the G-stabilizer of an s-dimensional totally isotropic subspace of $\mathbb{R}^{n,n}$, and is given by $N_\Phi A_\Phi M_{\Phi'}$ as follows.

$N_\Phi \cong \mathrm{Skew}\, \mathbb{R}^{s\times s} + \mathbb{R}^{s\times(n-s,n-s)} \cong \mathfrak{n}_\Phi$ as in (4.5.2).

$M_{\Phi'} A_\Phi \cong GL(s;\mathbb{R}) \times SO(n-s,n-s)$ and acts on N_Φ by $(\gamma,g)\colon (z,x) \mapsto (\gamma z \cdot {}^t\gamma, \gamma x \cdot {}^t g)$.

If $s = 1$ then N_Φ is abelian, $N_\Phi \cong \mathbb{R}^{n-1,n-1} \cong \mathfrak{n}_\Phi = \mathfrak{n}_{[\psi_1]}$, and $\mathfrak{m}_{\Phi'}$ acts on $\mathfrak{n}$ by $\underset{\psi_2}{\overset{1}{\circ}} - \cdots - \circ \lessdot \begin{smallmatrix} \circ\,\psi_{n-1} \\ \circ\,\psi_n \end{smallmatrix}$.

If $s = 2, 4,\ldots,2\left[\frac{n-2}{2}\right]$ then $\mathfrak{n}_\Phi$ is 2-step nilpotent with center $\mathfrak{n}_{2[\psi_s]} \cong \mathrm{Skew}\, \mathbb{R}^{s\times s}$, and with $\mathfrak{n}_{[\psi_s]} \cong \mathbb{R}^{s\times(n-s,n-s)}$. Then $\mathfrak{m}_{\Phi'}$ acts on $\mathfrak{n}_{2[\psi_s]}$ by $\underset{\psi_1}{\circ} - \underset{\psi_2}{\overset{1}{\circ}} - \cdots - \underset{\psi_{s-1}}{\circ} \otimes \underset{\psi_{s+1}}{\circ} - \cdots - \circ \lessdot \begin{smallmatrix} \circ\,\psi_{n-1} \\ \circ\,\psi_n \end{smallmatrix}$ if $s > 2$, $\circ \otimes \circ - \cdots - \circ \lessdot \begin{smallmatrix} \circ \\ \circ \end{smallmatrix}$ (trivially) if $s = 2$; and $\mathfrak{m}_{\Phi'}$ acts on $\mathfrak{n}_{[\psi_s]}$ by $\overset{1}{\circ} - \cdots - \circ \otimes \overset{1}{\circ} - \cdots - \circ \lessdot \begin{smallmatrix} \circ \\ \circ \end{smallmatrix}$.

(ii) $\Phi = \{\psi_{n-1}\}$ or $\Phi = \{\psi_n\}$. Then P_Φ is the G-stabilizer of a totally isotropic n-dimensional subspace of $\mathbb{R}^{n,n}$, $N_\Phi \cong \text{Skew}\,\mathbb{R}^{n\times n} \cong \mathfrak{n}_\Phi$, abelian, and $M_{\Phi'}A_\Phi \cong GL(n;\mathbb{R})$ acting on N_Φ and $\mathfrak{n}_\Phi$ by $\gamma: z \mapsto \gamma z \cdot {}^t\gamma$. So $\mathfrak{m}_{\Phi'}$ acts by $\underset{\psi_1}{\circ}\!-\!\!\!-\underset{\psi_2}{\overset{1}{\circ}}\!-\!\!\!-\cdots-\!\!\!-\underset{\psi_{n-2}}{\circ}\!-\!\!\!-\circ$. These two cases are $O(n,n)$-conjugate.

(iii) $\Phi = \{\psi_{n-1}, \psi_n\}$, n odd. Then P_Φ is the G-stabilizer of an (n-1)-dimensional totally isotropic subspace of $\mathbb{R}^{n,n}$ and $P_\Phi = N_\Phi A_\Phi M_{\Phi'}$ as follows. $N_\Phi \cong \text{Skew}\,\mathbb{R}^{(n-1)\times(n-1)} + \mathbb{R}^{(n-1)\times(1,1)}$ as in (4.5.2). $M_{\Phi'}A_\Phi \cong GL(n-1;\mathbb{R}) \times SO(1,1)$ and acts on N_Φ by $(\gamma,g):(z,x) \mapsto (\gamma z\cdot{}^t\gamma, \gamma x\cdot{}^t g)$. Here $\mathfrak{n}_\Phi$ is 2-step nilpotent with center $\mathfrak{n}_{[\psi_{n-1}]+[\psi_n]} \cong \text{Skew}\,\mathbb{R}^{(n-1)\times(n-1)}$, and with

$\mathfrak{n}_{[\psi_{n-1}]} \cong \mathbb{R}^{n-1} \cong \mathfrak{n}_{[\psi_n]}$. $\mathfrak{m}_{\Phi'}$ acts on $\mathfrak{n}_{[\psi_{n-1}]+[\psi_n]}$ by

$\underset{\psi_1}{\circ}\!-\!\!\!-\underset{\psi_2}{\overset{1}{\circ}}\!-\!\!\!-\cdots-\!\!\!-\underset{\psi_{n-2}}{\circ}$ and acts on each of $\mathfrak{n}_{[\psi_{n-1}]}$ and $\mathfrak{n}_{[\psi_n]}$ by

$\overset{1}{\circ}\!-\!\!\!-\circ\!-\!\!\!-\cdots-\!\!\!-\circ$.

§5. Passage to the General Classical Group

We use Theorem 2.1, Corollary 2.3 and a list of Satake diagrams to carry the classification of §4 over to arbitrary classical real simple Lie groups. The results of §§4 and 5 are summarized in §5.5 in matrix form.

5.1. Type A_{n-1}, $n \geqslant 2$. Here are the noncompact non-split cases.

5.1.1. $G = SL(n;\mathbb{C})$, complex form. In view of Theorem 2.1 we just complexify the results in (4.5.1). So (4.5.1)(i) gives the cases $\Phi = \{\psi_s\}$, $1 \leqslant s \leqslant n-1$, where

$$P_\Phi \cong \left\{ \begin{pmatrix} \alpha & x \\ 0 & \beta \end{pmatrix} : \begin{array}{l} \alpha \in GL(s;\mathbb{C}),\ \beta \in GL(n-s;\mathbb{C}), \\ (\det \alpha)(\det \beta) = 1,\ x \in \mathbb{C}^{s\times(n-s)} \end{array} \right\}$$

with $N_\Phi \cong \mathbb{C}^{s\times(n-s)} \cong \left\{\begin{pmatrix} I & x \\ 0 & I \end{pmatrix}\right\}$, abelian. And (4.5.1)(ii) gives the cases $\Phi = \{\psi_s, \psi_{n-s}\}$, $1 \leqslant s \leqslant \left[\frac{n-1}{2}\right]$, where

$$P_\Phi \cong \left\{ \begin{pmatrix} \alpha & x & z \\ 0 & \beta & y \\ 0 & 0 & \gamma \end{pmatrix} : \begin{array}{l} \alpha,\ \gamma \in GL(s;\mathbb{C}),\ \beta \in GL(n-2s;\mathbb{C}) \\ (\det \alpha)(\det \beta)(\det \gamma) = 1 \\ x \in \mathbb{C}^{s\times(n-2s)},\ y \in \mathbb{C}^{(n-2s)\times x},\ z \in \mathbb{C}^{s\times s} \end{array} \right\}$$

with $N_\Phi \cong \left\{\begin{pmatrix} I & x & z \\ 0 & I & y \\ 0 & 0 & I \end{pmatrix}\right\}$ and $Z_\Phi \cong \left\{\begin{pmatrix} I & 0 & z \\ 0 & I & 0 \\ 0 & 0 & I \end{pmatrix}\right\}$; so N_Φ is 2-step nilpotent and looks like a Heisenberg group.

5.1.2. $G = SL\left(\frac{n}{2};\mathbb{Q}\right)$, quaternion special linear group, real form with maximal compact subgroup $Sp(n/2)$. Here n is even and the Satake diagram is 

where we retain the root numbering of the Dynkin diagram. The simple $\mathfrak{a}$-roots are $\beta_s = \psi_{2s}|_{\mathfrak{a}}$. Here (4.5.1)(i) gives the cases $\Phi = \{\beta_s\}$, $1 \leqslant s \leqslant \frac{n-2}{2}$ corresponding to $\{\psi_{2s}\}$. Let $m = n/2$. Then

$$P_\Phi \cong \left\{ \begin{pmatrix} \alpha & x \\ 0 & \beta \end{pmatrix} : \begin{array}{l} \alpha \in GL(s;\mathbb{Q}),\ \beta \in GL(m-s;\mathbb{Q}), \\ (\det \alpha)(\det \beta) = 1,\quad x \in \mathbb{Q}^{s\times(m-s)} \end{array} \right\}$$

where $N_\Phi \cong \mathbb{Q}^{s\times(m-s)} \cong \left\{\begin{pmatrix} I & x \\ 0 & I \end{pmatrix}\right\}$, abelian. And (4.5.1)(ii) gives the cases $\Phi = \{\beta_s, \beta_{m-s}\}$, $1 \leqslant s \leqslant \left[\frac{n-1}{4}\right]$, corresponding to $\{\psi_{2s}, \psi_{n-2s}\}$. There

$$P_\Phi \cong \left\{ \begin{pmatrix} \alpha & x & z \\ 0 & \beta & y \\ 0 & 0 & \gamma \end{pmatrix} : \begin{array}{l} \alpha, \gamma \in GL(s;\mathbb{Q}),\ \beta \in GL(m-2s;\mathbb{Q}) \\ (\det \alpha)(\det \beta)(\det \gamma) = 1 \\ x \in \mathbb{Q}^{s\times(m-2s)},\ y \in \mathbb{Q}^{(m-2s)\times s},\ z \in \mathbb{Q}^{s\times s} \end{array} \right\}$$

with N_Φ 2-step nilpotent as above.

5.1.3. $G = SU(p,q)$, $p + q = n$, the special (determinant 1) unitary group of $\mathbb{C}^{p,q} = (\mathbb{C}^n, \langle\,,\rangle)$ where $\langle u,v\rangle = -\sum_1^p u_j\bar{v}_j + \sum_{p+1}^{p+q} u_k\bar{v}_k$. Here we may assume $1 \leqslant p \leqslant q$, so the Satake diagram is

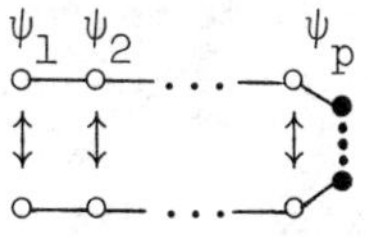

if $p < q$,
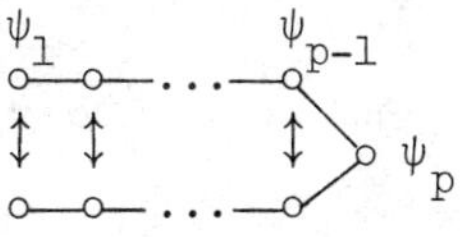

if $p = q$

The simple $\mathfrak{a}$-roots are the $\beta_s = \psi_s|_{\mathfrak{a}} = \psi_{n-s}|_{\mathfrak{a}}$ for $1 \leqslant s \leqslant p$. G does not have parabolic subgroups corresponding to the case (4.5.1)(i) except for $\Phi = \{\beta_s\}$ with $s = p = q$; and (4.5.1)(ii) gives the cases $\Phi = \{\beta_s\}, 1 \leqslant s \leqslant p$ and $s < q$, corresponding to $\{\psi_s, \psi_{n-s}\}$. P_Φ is the G-stabilizer of an s-dimensional totally isotropic subspace of $\mathbb{C}^{p,q}$, and is given by $N_\Phi A_\Phi M_\Phi$, as follows.

If $z \in \mathbb{C}^{s\times s}$ denote $\operatorname{Im}(z) = \frac{1}{2}(z - z^*)$, $z^* = {}^t\bar{z}$. So $\operatorname{Im} \mathbb{C}^{s\times s} = \{z \in \mathbb{C}^{s\times s} : z^* = -z\}$. Let $\mathbb{C}^{s\times(t,u)}$ denote $\mathbb{C}^{s\times t} \oplus \mathbb{C}^{s\times u}$ with $\mathcal{H}: \mathbb{C}^{s\times(t,u)} \times \mathbb{C}^{s\times(t,u)} \to \mathbb{C}^{s\times s}$, hermitian map given by $\mathcal{H}((x_1,x_2),(y_1,y_2)) = x_1y_1^* - x_2y_2^*$. Then $N_\Phi \cong \operatorname{Im} \mathbb{C}^{s\times s} + \mathbb{C}^{s\times(p-s,q-s)} \cong \mathfrak{n}_\Phi$ with

$$(z,x)(z',x') = (z,z'+\operatorname{Im}\mathcal{H}(x,x'),x+x') \quad \text{and} \quad [(z,x),(z',x')] = (2\operatorname{Im}\mathcal{H}(x,x'),0).$$

Further, $M_\Phi A_\Phi \cong GL(s;\mathbb{C}) \times SU(p-s,q-s)$, acting on N_Φ and $\mathfrak{n}_\Phi$ by $(\gamma,g): (z,x) \mapsto (\gamma z\gamma^*, \gamma x g^*)$.

If $p = q = s$ then $N_\Phi \cong \operatorname{Im} \mathbb{C}^{s\times s}$, abelian. Otherwise, N_Φ is 2-step nilpotent with $\mathfrak{z}_\Phi = \mathfrak{n}_{2[\beta_s]} \cong \operatorname{Im} \mathbb{C}^{s\times s}$ and $\mathfrak{n}_{[\beta_s]} \cong \mathbb{C}^{s\times(p-s,q-s)}$.

5.2. Type B_n, $n \geqslant 2$. Here are the noncompact non-split cases.

5.2.1. $G = SO(2n+1;\mathbb{C})$, complex form. In view of Theorem 2.1 we just complexify the result of (4.5.2). So $\mathbb{C}^{2n+1}$ is endowed with a nondegenerate symmetric bilinear form, G consists of the elements of determinant 1 in its orthogonal group, and we have the cases $\Phi = \{\psi_s\}$ with $s = 1$ or $s = 2,4,\ldots,2\left[\frac{n}{2}\right]$. P_Φ then is the stabilizer of an s-dimensional totally isotropic subspace, given by $N_\Phi A_\Phi M_\Phi$, where

$$N_\Phi \cong \text{Skew } \mathbb{C}^{s\times s} + \mathbb{C}^{s\times(2n+1-2s)}$$

with $(z,x)(z',x') = (z+z' + (x,x'),x+x')$ as in (4.5.2), and $M_\Phi A_\Phi \cong GL(s;\mathbb{C}) \times SO(2n+1-2s)$ acting by $(\gamma,g)\colon (z,x) \mapsto (\gamma z\cdot{}^t\gamma, \gamma x\cdot{}^t g)$.

5.2.2. $G = SO(p,q)$, $1 \leqslant p \leqslant q$, $p + q = 2n+1$. The Satake diagram is $\overset{\psi_1}{\circ}-\cdots-\overset{\psi_p}{\circ}-\bullet-\cdots-\bullet\Rightarrow\bullet$ for $p \leqslant n-1$, $\overset{\psi_1}{\circ}-\cdots-\circ\Rightarrow\overset{\psi_n}{\circ}$ for $p = n$. The latter is the split case, already studied. The simple $\mathfrak{a}$-roots are the $\beta_s = \psi_s|_{\mathfrak{a}}$, $1 \leqslant s \leqslant p$, and (4.5.2) gives us the cases $\Phi = \{\beta_s\}$ with $s = 1$ or $s = 2,4,\ldots,2\left[\frac{p}{2}\right]$. Then P_Φ is the G-stabilizer of an s-dimensional totally isotropic subspace of $\mathbb{R}^{p,q}$. It is given by $N_\Phi A_\Phi M_\Phi$, where

$$N_\Phi \cong \text{Skew } \mathbb{R}^{s\times s} + \mathbb{R}^{s\times(p-s,q-s)} \quad \text{as in (4.5.2)},$$

$$M_\Phi A_\Phi \cong GL(s;\mathbb{R}) \times SO(p-s,q-s),$$

and $M_\Phi A_\Phi$ acts by $(\gamma,g)\colon(z,x) \mapsto (\gamma a\cdot{}^t\gamma, \gamma x\cdot{}^t g)$ as before.

If $s = 1$ then $N_\Phi \cong \mathbb{R}^{p-s,q-s}$, abelian.

If $s = 2,4,\ldots,2[p/2]$ then $\mathfrak{n}_\Phi$ is 2-step nilpotent with center $\mathfrak{n}_{2[\beta_s]} \cong \text{Skew } \mathbb{R}^{s\times s}$ and with $\mathfrak{n}_{[\beta_s]} \cong \mathbb{R}^{s\times(p-s,q-s)}$.

5.3. Type C_n, $n \geqslant 2$. Here are the noncompact non-split cases.

5.3.1. $G = Sp(n;\mathbb{C})$ complex form. We just complexify the results in (4.5.3). G is the group of all linear transformations of $\mathbb{C}^{2n}$ that preserve a nondegenerate antisymmetric bilinear form $\{\ ,\ \}$, (4.5.3) gives the cases $\Phi = \{\psi_s\}$, $1 \leqslant s \leqslant n$, and P_Φ is the G-stabilizer of an s-dimensional totally isotropic subspace of $(\mathbb{C}^{2n}, \{,\})$. Here $P_\Phi = N_\Phi A_\Phi M_\Phi$, where $N_\Phi \cong \text{Sym}\, \mathbb{C}^{s\times s} + \mathbb{C}^{s\times 2(n-2)} \cong \mathfrak{n}_\Phi$ with composition as in (4.5.3), and $M_\Phi A_\Phi \cong GL(s;\mathbb{C}) \times Sp(n-s;\mathbb{C})$ acting by $(\gamma,g): (z,x) \mapsto (\gamma z \cdot {}^t\gamma, \gamma x \cdot {}^t g)$.

If $s = n$ then $N_\Phi \cong \text{Sym}\, \mathbb{C}^{n\times n} \cong \mathfrak{n}_\Phi$, abelian.

If $1 \leqslant s < n$ then $\mathfrak{n}_\Phi$ is 2-step nilpotent with center $\mathfrak{n}_{2[\psi_s]} \cong \text{Sym}\, \mathbb{C}^{s\times s}$ and with $\mathfrak{n}_{[\psi_s]} \cong \mathbb{C}^{s\times 2(n-s)}$.

5.3.2. $G = Sp(p,q)$, $1 \leqslant p \leqslant q$, $p + q = n$, the unitary group of $\mathbb{Q}^{p,q} = (\mathbb{Q}^n, \langle\, ,\rangle)$ where $\langle u,v \rangle = - \sum_1^p u_j \bar{v}_j + \sum_{p+1}^{p+q} u_j \bar{v}_j$. The Satake diagram is

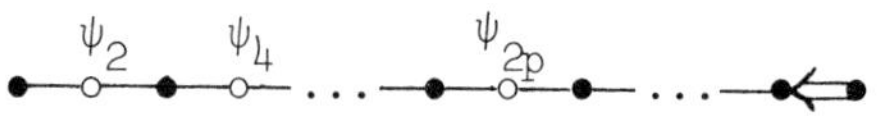

for $p < q$,

for $p = q$.

The simple $\mathfrak{a}$-roots are the $\beta_s = \psi_{2s}|_{\mathfrak{a}}$, $1 \leqslant s \leqslant p$, and (4.5.3) gives the cases $\Phi = \{\beta_s\}$, $1 \leqslant s \leqslant p$, corresponding to $\{\psi_{2s}\}$. P_Φ is the G-stabilizer of an s-dimensional (over $\mathbb{Q}$) totally isotropic subspace of $\mathbb{Q}^{p,q}$, and is given by $N_\Phi A_\Phi M_\Phi$, as follows.

$N_\Phi \cong \text{Im}\, \mathbb{Q}^{s\times s} + \mathbb{Q}^{s\times(p-s,q-s)} \cong \mathfrak{n}_\Phi$ as in (5.1.3) where we replace $\mathbb{C}$ by $\mathbb{Q}$. $M_\Phi A_\Phi \cong GL(s;\mathbb{Q}) \times Sp(p-s,q-s)$, acting on N_Φ and $\mathfrak{n}_\Phi$ by $(\gamma,g): (z,x) \mapsto (\gamma z \gamma^*, \gamma x g^*)$.

If $p = q = s$ then $N_\Phi \cong \text{Im}\, \mathbb{Q}^{s\times s}$, abelian. Otherwise, N_Φ is 2-step nilpotent with $\mathfrak{z}_\Phi = \mathfrak{n}_{2[\beta_s]} \cong \text{Im}\, \mathbb{Q}^{s\times s}$ and $\mathfrak{n}_{[\beta_s]} \cong \mathbb{Q}^{s\times(p-s,q-s)}$.

5.4. Type D_n, $n \geq 4$. Here are the noncompact non-split cases.

5.4.1. $G = SO(2n;\mathbb{C})$, complex form. We complexify (4.5.4). G consists of the elements of determinant 1 in the orthogonal group of a $\mathbb{C}^{2n}$ with nondegenerate symmetric bilinear form. From (4.5.4)(i) we have the cases $\Phi = \{\psi_s\}$ with $s = 1$ and with $s = 2,4,\ldots,2\left[\frac{n-2}{2}\right]$. There $N_\Phi \cong \text{Skew } \mathbb{C}^{s\times s} + \mathbb{C}^{s\times 2(n-s)} \cong \mathfrak{n}_\Phi$ with $(z,x)(z',x') = (z+z' + \delta(x,x'),\ x+x')$ as in (4.5.2), and $M_\Phi A_\Phi \cong GL(s;\mathbb{C}) \times SO(2(n-s);\mathbb{C})$ acting by $(\gamma,g):(z,x) \mapsto (\gamma z\cdot {}^t\gamma, \gamma x\cdot {}^t g)$. If $s = 1$ then $N_\Phi \cong \mathbb{C}^{2n-2}$, abelian; otherwise N_Φ is 2-step nilpotent with $\mathfrak{z}_\Phi = \mathfrak{n}_{2[\psi_s]} \cong \text{Skew } \mathbb{C}^{s\times s}$ and with $\mathfrak{n}_{[\psi_s]} \cong \mathbb{C}^{s\times 2(n-s)}$.

From (4.5.4)(ii) we have the cases $\Phi = \{\psi_{n-1}\}$ and $\Phi = \{\psi_n\}$. Those P_Φ are isomorphic, related by an outer automorphism of G. Each is the G-stabilizer of a totally isotropic n-dimensional subspace of $\mathbb{C}^{2n}$, with $N_\Phi \cong \text{Skew } \mathbb{C}^{n\times n} \cong \mathfrak{n}_\Phi$, abelian, and with $M_\Phi A_\Phi \cong GL(n;\mathbb{C})$ acting by $\gamma: z \mapsto \gamma z\cdot {}^t\gamma$.

If n is odd then we also have, from (4.5.4)(iii), the case $\Phi = \{\psi_{n-1},\psi_n\}$, where P_Φ is the G-stabilizer of an $(n-1)$-dimensional totally isotropic subspace of $\mathbb{C}^{2n}$. There $N_\Phi \cong \text{Skew } \mathbb{C}^{(n-1)\times(n-1)} + \mathbb{C}^{(n-1)\times 2} \cong \mathfrak{n}_\Phi$, as above, with $M_\Phi A_\Phi \cong GL(n-1;\mathbb{C}) \times SO(2;\mathbb{C})$ acting by $(\gamma,g): (z,x) \mapsto (\gamma z\cdot {}^t\gamma, \gamma x\cdot {}^t g)$. N_Φ is 2-step nilpotent with $\mathfrak{z}_\Phi = \mathfrak{n}_{2[\psi_{n-1} \text{ or } n]} \cong \mathbb{C}^{(n-1)\times(n-1)}$ and with $\mathfrak{n}_{[\psi_{n-1} \text{ or } n]} \cong \mathbb{C}^{(n-1)\times 2}$.

5.4.2. $G = SO(p,q)$, $1 \leq p \leq q$, $p + q = 2n$. The Satake diagram is

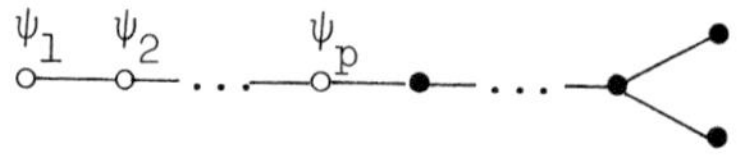

if $p \leqslant n-2$,

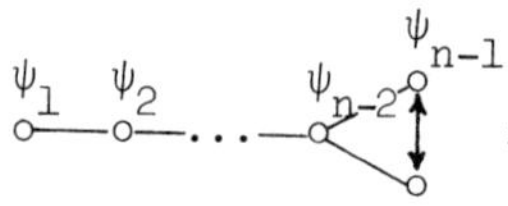

if $p = n-1$,

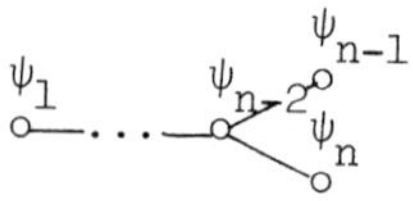

if $p = n$.

The last of these is the split case, already studied. The simple $\mathfrak{a}$-roots are the $\beta_s = \psi_s|_{\mathfrak{a}}$, $1 \leqslant s \leqslant p$.

From (4.5.4)(i) we have the cases $\Phi = \{\beta_s\}$ with $s = 1$ and with $s = 2,4,\ldots,\min\left(2\left[\frac{p}{2}\right], 2\left[\frac{n-2}{2}\right]\right)$. Then P_Φ is the G-stabilizer of an s-dimensional totally isotropic subspace of $\mathbb{R}^{p,q}$, given by $N_\Phi \cong \text{Skew}\,\mathbb{R}^{s\times s} + \mathbb{R}^{s\times(p-s,q-s)}$ as in (4.5.2), and by $M_\Phi A_\Phi \cong GL(s;\mathbb{R}) \times SO(p-s,q-s)$ acting as $(\gamma,g):(z,x) \mapsto (\gamma z\cdot{}^t\gamma, \gamma x\cdot{}^tg)$. N_Φ is abelian when $s = 1$, 2-step nilpotent when $s = 2,4,\ldots,\min\left(2\left[\frac{p}{2}\right],2\left[\frac{n-2}{2}\right]\right)$.

A parabolic subgroup of G is specified by (4.5.4)(ii) only when $p = n = q$, where G is the split D_n. Those

$$P_{\{\beta_{n-1}\}} \cong P_{\{\beta_n\}} \cong \text{Skew}\,\mathbb{R}^{n\times n} \cdot GL(n;\mathbb{R})$$

were, of course, discussed in (4.5.4)(ii).

If n is odd and $n-1 \leqslant p \leqslant n$, then (4.5.4)(iii) gives the cases $\Phi = \{\beta_{n-1}\}$ for $p = n-1$, $\Phi = \{\beta_{n-1},\beta_n\}$ for $p = n$. There P_Φ is the G-stabilizer of a totally isotropic (n-1)-dimensional subspace of $\mathbb{R}^{p,q}$,

$$P_\Phi \cong \{\text{Skew}\,\mathbb{R}^{(n-1)\times(n-1)} + \mathbb{R}^{(n-1)\times(1,1)}\}\cdot\{GL(n-1;\mathbb{R}) \times SO(1,1)\}$$

as above, with N_Φ 2-step nilpotent.

5.4.3. $G = SO^*(2n)$, real group of type D_n with maximal compact subgroup $U(n)$, isomorphic to the group of all linear transformations of $\mathbb{Q}^n$ that preserve a nondegenerate skew-hermitian form $[u,v] = \sum u_a i \bar{v}_a$. The Satake diagram is

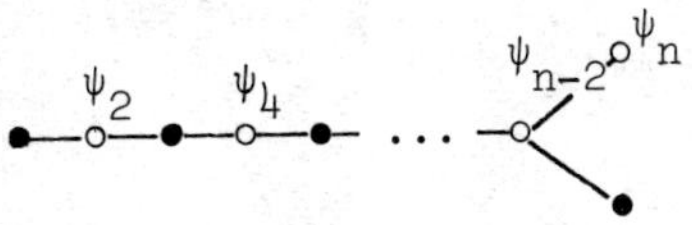

for n even

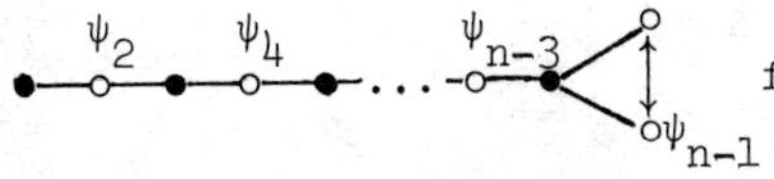

for n odd

The simple $\mathfrak{a}$-roots are the $\beta_s = \psi_{2s}|_{\mathfrak{a}}$, $1 \leqslant s \leqslant [n/2]$.

From (4.5.4), parts (i), (ii) and (iii), we have the cases $\Phi = \{\beta_s\}$, $1 \leqslant s \leqslant [n/2]$. There P_Φ is the G-stabilizer of a s-dimensional totally isotropic subspace of $(\mathbb{Q}^n, [\ ,\ ])$. If one translates the complex matrix description [28, Theorem 13.9] of P_Φ to a quaternion matrix description along the ideas of [30, §8], he sees that $P_\Phi = N_\Phi A_\Phi M_\Phi$, where

$$N_\Phi \cong \mathfrak{so}^*(2s) + \mathbb{Q}^{s\times(n-2s)} \cong \mathfrak{n}_\Phi ,$$

with

$$(z,x)(z',x') = (z+z' + \tfrac{1}{2}\{xix'^*i - x'ix^*i\} , x+x')$$

$$[(z,x),(z',x')] = (xix'^*i - x'ix^*i,0) ,$$

and with

$$M_\Phi A_\Phi = GL(s;\mathbb{Q}) \times SO^*(2(n-2s)) .$$

For the action of $M_\Phi A_\Phi$ on N_Φ, first note that

$$SO^*(2r) = \{g \in \mathbb{Q}^{r\times r} : gig^* = iI\}$$

so

$$\mathfrak{so}^*(2r) = \{z \in \mathbb{Q}^{r\times r} : zi + iz^* = 0\} = \{z \in \mathbb{C}^{r\times r} : (zi)^* = zi\} .$$

Now $(\gamma,g) \in GL(s;\mathbb{Q}) \times SO^*(2(n-2s))$ acts on N_Φ by $zi \mapsto \gamma\cdot zi\cdot\gamma^*$ and $x \mapsto \gamma x g^*$. More formally,

$$(\gamma,g):(z,x) \mapsto (-\gamma zi\gamma^* i, \gamma x g^*).$$

If $n = 2s$ then $N_\Phi \cong \mathfrak{so}^*(n) \cong \mathfrak{n}_\Phi$, abelian.

If $n > 2s$ then N_Φ is 2-step nilpotent with $\mathfrak{z}_\Phi = \mathfrak{n}_{2[\beta_s]} \cong \mathfrak{so}^*(2s)$ and $\mathfrak{n}_{[\beta_s]} \cong \mathbb{Q}^{s\times(n-2s)}$.

5.5. Summary. The parabolic subgroups $P \subset G$, where G is a classical simple Lie group and the nilradical of P has square integrable representations, are given up to conjugacy as follows for a choice of G within its local isomorphism class. Here it is convenient to put the $SO(p,q)$, $SU(p,q)$ and $Sp(p,q)$ into one family, the $SU(p,q;\mathbb{F})$ as $\mathbb{F} = \mathbb{R}, \mathbb{C}, \mathbb{Q}$, and to use matrix formulation.

$SL(n;\mathbb{F})$, $\mathbb{F} = \mathbb{R}$, $\mathbb{C}$ or $\mathbb{Q}$, $n \geqslant 2$.

(i) $N_\Phi \cong \mathbb{F}^{s\times s}$, commutative; $1 \leqslant s \leqslant n-1$.

$$M_\Phi A_\Phi \cong \left\{\begin{pmatrix}\alpha & 0\\ 0 & \beta\end{pmatrix} : \alpha \in GL(s;\mathbb{F}),\ \beta \in GL(n-s;\mathbb{F}),\ \det\begin{pmatrix}\alpha & 0\\ 0 & \beta\end{pmatrix} = 1\right\}$$

acts on N_Φ by $z \mapsto \alpha z \beta^{-1}$

(ii) $N_\Phi \cong \mathbb{F}^{s\times s} + \mathbb{F}^{s\times(n-2s)} + \mathbb{F}^{(n-2s)\times s}$, 2-step nilpotent;

$1 \leqslant s \leqslant \left[\frac{n-1}{2}\right]$.

Composition: $(z,x,y)(z',x',y') = (z+z' + xy', x+x', y+y')$

$$M_\Phi A_\Phi \cong \left\{\begin{pmatrix}\alpha & 0 & 0\\ 0 & \beta & 0\\ 0 & 0 & \gamma\end{pmatrix} : \begin{array}{l}\alpha,\gamma \in GL(s;\mathbb{F}),\ \beta \in GL(n-2s;\mathbb{F})\\ (\det\alpha)(\det\beta)(\det\gamma) = 1\end{array}\right\}$$

acts on N_Φ by $(z,x,y) \mapsto (\alpha z\gamma^{-1}, \alpha x\beta^{-1}, \beta y\gamma^{-1})$

$SU(p,q;\mathbb{F})$, $\mathbb{F} = \mathbb{R}$, $\mathbb{C}$ or $\mathbb{Q}$, $1 \leqslant p \leqslant q$, $p + q \geqslant 3$ in real case.

$N_\Phi \cong \operatorname{Im} \mathbb{F}^{s\times s} + \mathbb{F}^{s\times(p-s,q-s)}$ where $1 \leqslant s \leqslant p$,

and where in the real case either $s = 1$ or $p = q = s$

or s is even.

Composition: $(z,x)(z',x') = (z+z' + \operatorname{Im}\mathcal{H}(x,x'), x+x')$.

2-step nilpotent except in the abelian cases

$p = q = s$ (any $\mathbb{F}$) and $s = 1$, $\mathbb{F} = \mathbb{R}$

$M_\Phi A_\Phi \cong GL(s;\mathbb{F}) \times SU(p-s,q-s;\mathbb{F})$, acts on

N_Φ by $(\gamma,g): (z,x) \mapsto (\gamma z\gamma^*, \gamma x g^*)$.

$SO(n;\mathbb{C})$, $n \geqslant 3$.

$N_\Phi \cong \text{Skew } \mathbb{C}^{s\times s} + \mathbb{C}^{s\times(n-2s)}$ where

either $s = 1$ or $2s = n$ (commutative cases)

or $s = 2,4,\ldots,2\left[\frac{n-1}{2}\right]$ (2-step nilpotent cases).

Composition: $(z,x)(z',x') = (z+z' + \mathcal{S}(x,x'), x+x')$.

$M_\Phi A_\Phi \cong GL(s;\mathbb{C}) \times SO(n-2s;\mathbb{C})$, acts on N_Φ by

$(\gamma,g):(z,x) \mapsto (\gamma z\cdot {}^t\gamma, \gamma x\cdot {}^t g)$.

$Sp(n;\mathbb{F})$, $\mathbb{F} = \mathbb{R}$ or $\mathbb{C}$, $n \geqslant 2$.

$N_\Phi \cong \text{Sym}\,\mathbb{F}^{s\times s} + \mathbb{F}^{s\times 2(n-s)}$, $1 \leqslant s \leqslant n$, commutative when $s = n$ and 2-step nilpotent when $1 \leqslant s < n$.

Composition: $(z,x)(z',x') = (z+z' + \langle x,x'\rangle, x+x')$.

$M_\Phi A_\Phi \cong GL(s;\mathbb{F}) \times Sp(n-s;\mathbb{F})$, acts on N_Φ by $(\gamma,g):(z,x) \mapsto (\gamma z\cdot{}^t\gamma\,, \gamma x\cdot{}^t g)$

$SO^*(2n) = \{g \in \mathbb{Q}^{n\times n}: gig^* = i\}$, $n \geqslant 3$.

$N_\Phi \cong \mathfrak{so}^*(2s) + \mathbb{Q}^{s\times(n-2s)}$, $1 \leqslant s \leqslant [n/2]$,

commutative if $2s = n$, 2-step nilpotent otherwise.

Composition: $(z,x)(z',x') = (z+z' + \frac{1}{2}(xix'^* - x'ix^*), x+x')$.

$M_\Phi A_\Phi \cong GL(s;\mathbb{Q}) \times SO^*(2n-4s)$, acts on N_Φ by

$(\gamma,g):(z,x) \to (\gamma z\gamma^*, \gamma x g^*)$.

If $\tilde{G}$ is a reductive Lie group whose semisimple part $\tilde{G}'$ is locally isomorphic to one of the groups G listed above, then the parabolics $\tilde{P} \subset \tilde{G}$ whose nilradicals have square integrable representations are just the ones corresponding to the listed parabolics $P \subset G$.

§6. Classification in the Real Split Exceptional Groups

We run through the list of real split simple exceptional groups and find all parabolic subgroups whose nilradical has square integrable representations. As before, results from §3 are used to limit the possibilities. This time, however, we cannot fall back on results of [28] to exhibit square integrable representations in the cases that survive the screening. Instead, either we recognize the nilradical as being a commutative or Heisenberg group, or we use a partition argument as in (4.1.2), or in some case a special argument.

6.1. Split G_2: $\overset{-\mu}{\circ}$---$\overset{\psi_1}{\circ}\Rrightarrow\overset{\psi_2}{\circ}$. Write (m,n) for a root $m\psi_1 + n\psi_2$. Then the positive root system is

$$\Delta^+ = \{(1,0),\ (0,1),\ (1,1),\ (1,2),\ (1,3),\ (2,3)\} .$$

Lemma 3.9 leaves only the possibilities $\{\psi_1\}$ and $\{\psi_2\}$ for Φ .

(i) $\Phi = \{\psi_1\}$. Then $\Delta_\Phi^+ = \{[\psi_1],\ 2[\psi_1]\}$, $[\psi_1]$ consists of (1,0), (1,1), (1,2) and (1,3), and $2[\psi_1]$ consists just of (2,3). As $(1,0) + (1,3) = (2,3) = (1,1) + (1,2)$, $\mathfrak{n}_\Phi$ is the 5-dimensional Heisenberg algebra, so N_Φ has square integrable representations.

(ii) $\Phi = \{\psi_2\}$. Then $\Delta_\Phi^+ = \{[\psi_2], 2[\psi_2],\ 3[\psi_2]\}$, $[\psi_2] = \{(0,1),\ (1,1)\}$ so $\dim \mathfrak{n}_{[\psi_2]} = 2$, and $2[\psi_2] = \{(1,2)\}$ so $\dim \mathfrak{n}_{2[\psi_2]} = 1$. Lemma 3.8 says that N_Φ cannot have square integrable representations.

6.2. Split F_4: 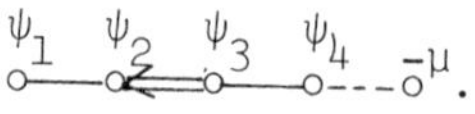

. Write $k\ell mn$ for a root $k\psi_1 + \ell_2\psi + m\psi_3 + n\psi_4$. Then the positive roots are

1000	0100	0010	0001	1100	0110
0011	1110	0210	0111	1210	1111
0211	1211	2210	0221	2211	1221
2221	1321	2321	2421	2431	2432

Lemma 3.9 says that Φ has just one element.

(i) $\Phi = \{\psi_1\}$. Then $\Delta_\Phi^+ = \{[\psi_1], 2[\psi_1]\}$ and we enumerate

$[\psi_1]$: 1000, 1100, 1110, 1210, 1111, 1211, 1221, 1321.
$2[\psi_1]$: 2210, 2211, 2221, 2321, 2421, 2431, 2432.

Let $0 \neq f \in \mathfrak{n}_\Phi^*$ with $f(\mathfrak{g}_\alpha) = 0$ for $\alpha \neq 2321$. Since

$$1210 + 1111 = 1110 + 1211 = 1100 + 1221 = 1000 + 1321 = 2321$$

b_f is nonsingular on $\mathfrak{n}_{[\psi_1]} \approx \mathfrak{n}_\Phi/\mathfrak{z}_\Phi$. So N_Φ has square integrable representations.

(ii) $\Phi = \{\psi_2\}$. Then $\Delta_\Phi^+ = \{[\psi_2], 2[\psi_2], 3[\psi_2], 4[\psi_2]\}$. Note that $[\psi_2]$ consists of 6 roots while $3[\psi_2]$ consists of 2, so $\dim \mathfrak{n}_{[\psi_2]} \neq \dim \mathfrak{n}_{3[\psi_2]}$. Lemma 3.8 says that N_Φ cannot have square integrable representations

(iii) $\Phi = \{\psi_3\}$. Then $\Delta_\Phi^+ = \{[\psi_3], 2[\psi_3], 3[\psi_3]\}$. Here $[\psi_3]$ consists of 12 roots and $2[\psi_3]$ consists of 6, so $\dim \mathfrak{n}_{[\psi_3]} \neq \dim \mathfrak{n}_{2[\psi_3]}$. Lemma 3.8 says that N_Φ cannot have square integrable representations.

(iv) $\Phi = \{\psi_4\}$. Then $\Delta_\Phi^+ = \{[\psi_4], 2[\psi_4]\}$ where $[\psi_4]$ consists of 14 roots, $2[\psi_4]$ consists just of the highest root $\mu = 2432$, and no root in $[\psi_4]$ is orthogonal to μ. Thus $\mathfrak{n}_\Phi$ is the 15-dimensional Heisenberg algebra and so N_Φ has square integrable representations.

<u>6.3. Split</u> E_6: 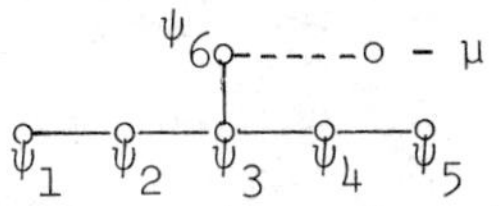

. Write $ijk\ell mn$ for a root $i\psi_1 + j\psi_2 + k\psi_3 + \ell\psi_4 + m\psi_5 + n\psi_6$. Then the positive roots are

100000	010000	001000	000100	000010	000001
110000	011000	001100	000110	001001	111000
011100	001110	011001	001101	111100	011110
001111	011101	111001	111110	011111	111101
111111	012101	012111	112101	012211	122101
112111	112211	122111	122211	123211	123212

Lemma 3.9 excludes all possibilities of Φ with more than one element, except the $\{\psi_1$ or ψ_2, ψ_4 or $\psi_5\}$.

(i) $\Phi = \{\psi_1\}$. Then $\Delta_\Phi^+ = \{[\psi_1]\}$, so $\mathfrak{n}_\Phi$ is a 16-dimensional commutative algebra and N_Φ has square integrable representations.

(ii) $\Phi = \{\psi_2\}$. Then $\Delta_\Phi^+ = \{[\psi_2], 2[\psi_2]\}$, and $\mathfrak{m}_{\Phi'}$ acts on the center $\mathfrak{z}_\Phi = \mathfrak{n}_{2[\psi_2]}$ by 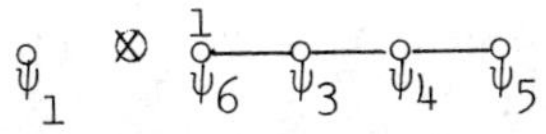

because μ is the highest weight vector there. Now the identity component $(M_{\Phi'})_0$ has an open orbit on $\mathfrak{z}_\Phi^*$, so it does not fix any nonzero polynomial there. Were N_Φ to have square integrable representations, the Pffafian polynomial function on $\mathfrak{z}_\Phi^*$ would be $(M_{\Phi'})_0$ - invariant, for any automorphism γ of $\mathfrak{n}_\Phi$ would multiply it by $\det(\gamma|_{\mathfrak{z}_\Phi})$. Thus N_Φ does not have square integrable representations.

(iii) $\Phi = \{\psi_3\}$. Then $\Delta_\Phi^+ = \{[\psi_3], 2[\psi_3], 3[\psi_3]\}$, $[\psi_3]$ consists of 18 roots, and $2[\psi_3]$ consists of 9 roots, so $\dim \mathfrak{n}_{[\psi_3]} \neq \dim \mathfrak{n}_{2[\psi_3]}$. Lemma 3.8 says that N_Φ cannot have square integrable representations.

(iv) $\Phi = \{\psi_4\}$. As in (ii), N_Φ does not have square integrable representations.

(v) $\Phi = \{\psi_5\}$. As in (i), N_Φ is a 16-dimensional commutative group and has square integrable representations.

(vi) $\Phi = \{\psi_6\}$. Then $\Delta_\Phi^+ = \{[\psi_6], 2[\psi_6]\}$, $2[\psi_6]$ consists of the single root 123212, and $[\psi_6]$ consists of 20 roots falling into 10 pairs with sum 123212:

000001	and	123211,	001001	and	122211,
011001	and	112211,	001101	and	122111,
001111	and	122101,	011101	and	112111,
111001	and	012211,	011111	and	112101,
111101	and	012111,	111111	and	012101.

Now N_Φ is the 21-dimensional Heisenberg group, so it has square integrable representations.

(vii) $\Phi = \{\psi_2, \psi_4\}$. Then $\Delta_\Phi^+ = \{[\psi_2], [\psi_4], [\psi_2] + [\psi_4], 2[\psi_2] + [\psi_4], [\psi_2] + 2[\psi_4], 2[\psi_2] + 2[\psi_4]$. Here $[\psi_2]$ consists of the 6 roots 010000, 110000, 011000, 111000, 011001 and 111001, while $[\psi_2] + 2[\psi_4]$ consists of the 2 roots 012211 and 112211, so $\dim \mathfrak{n}_{[\psi_2]} \neq \dim \mathfrak{n}_{[\psi_2] + 2[\psi_4]}$. Lemma 3.8 says that N_Φ cannot have square integrable representations.

(viii) $\Phi = \{\psi_2,\psi_5\}$. Then $\Delta_\Phi^+ = \{[\psi_2], [\psi_5], 2[\psi_2], [\psi_2] + [\psi_5], 2[\psi_2] + [\psi_5]\}$, $[\psi_5]$ consists of the 4 roots 000010, 000110, 001110, 001111, and $2[\psi_2]$ consists of the one root 122101, so $\dim \mathfrak{n}_{[\psi_5]} \neq \dim \mathfrak{n}_{2[\psi_2]}$. Lemma 3.8 says that N_Φ cannot have square integrable representations.

(ix) $\Phi = \{\psi_1,\psi_4\}$. As in (viii), N_Φ does not have square integrable representations.

(x) $\Phi = \{\psi_1,\psi_5\}$. Then $\Delta_\Phi^+ = \{[\psi_1], [\psi_5], [\psi_1] + [\psi_5]\}$. Partition $[\psi_1] = S_1 \cup S_2$ and $[\psi_5] = T_1 \cup T_2$ where

S_1: $\xi_1 = 100000$, $\xi_2 = 111100$, $\xi_3 = 111101$, $\xi_4 = 112101$

S_2: $\xi_5 = 110000$, $\xi_6 = 111000$, $\xi_7 = 111001$, $\xi_8 = 122101$

T_1: $\eta_1 = 012211$, $\eta_2 = 001111$, $\eta_3 = 001110$, $\eta_4 = 000110$

T_2: $\eta_5 = 012111$, $\eta_6 = 011111$, $\eta_7 = 011110$, $\eta_8 = 000010$

Then no $\xi_i + \xi_j$ and no $\eta_i + \eta_j$ is a root, $\xi_i + \eta_i = 112211 \in [\psi_1] + [\psi_5]$ for $1 \leqslant i \leqslant 4$, $\xi_i + \eta_i = 122111 \in [\psi_1] + [\psi_5]$ for $5 \leqslant i \leqslant 8$, and no other sums $\xi_i + \eta_j$ give 112211 or 122111. Choose nonzero linear functionals f_1, f_2 on $\mathfrak{n}_\Phi$ such that $f_1(\mathfrak{g}_\alpha) = 0$ for $\alpha \neq 112211$ and $f_2(\mathfrak{g}_\beta) = 0$ for $\beta \neq 122111$. Then $f = f_1 + f_2$ has the property that $b_f(x,y) = f[x,y]$ is nonsingular on $\mathfrak{n}_{[\psi_1]} + \mathfrak{n}_{[\psi_5]} \approx \mathfrak{n}_\Phi/\mathfrak{z}_\Phi$, so N_Φ has square integrable representations.

What is N_Φ ? Enumerate $[\psi_1] + [\psi_5]$; it consists of $\zeta_1 = 123212$, $\zeta_2 = 123211$, $\zeta_3 = 122211$, $\zeta_4 = 122111$, $\zeta_5 = 112211$, $\zeta_6 = 112111$, $\zeta_7 = 111111$ and $\zeta_8 = 111110$. The multiplication table of $\mathfrak{n}_\Phi$ is specified, up to nonzero structure constants, by knowing just which $\xi_i + \eta_j = \zeta_k$. We summarize that:

	ξ_1	ξ_2	ξ_3	ξ_4	ξ_5	ξ_6	ξ_7	ξ_8
η_1	ζ_5	-	-	-	ζ_3	ζ_2	ζ_1	-
η_2	-	ζ_5	-	-	ζ_7	ζ_6	-	ζ_1
η_3	-	-	ζ_5	-	ζ_8	-	ζ_6	ζ_2
η_4	-	-	-	ζ_5	-	ζ_8	ζ_7	ζ_3
η_5	ζ_6	ζ_2	ζ_1	-	ζ_4	-	-	-
η_6	ζ_7	ζ_3	-	ζ_1	-	ζ_4	-	-
η_7	ζ_8	-	ζ_3	ζ_2	-	-	ζ_4	-
η_8	-	ζ_8	ζ_7	ζ_6	-	-	-	ζ_4

Conclusion: $N_\Phi \cong \left\{\begin{pmatrix}1 & x & z\\ 0 & 1 & y\\ 0 & 0 & 1\end{pmatrix} : x, y, z \in \mathfrak{C}_{\mathbb{R}}\right\}$ where $\mathfrak{C}_{\mathbb{R}}$ is the split real Cayley algebra.

6.4. Split E_7: 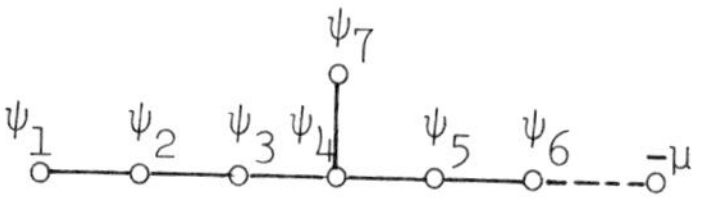

. Write hijkℓmn for a root $h\psi_1 + i\psi_2 + j\psi_3 + k\psi_4 + \ell\psi_5 + m\psi_6 + n\psi_7$. Then the positive roots are (i) the ones listed for E_6, each preceded by a zero, and (ii) the roots labeled

1000000	1100000	1110000	1111000	1111100
1111001	1111110	1111101	1111111	1112101
1112111	1122101	1112211	1122111	1222101
1122211	1222111	1123211	1222211	1123212
1223211	1223212	1233211	1233212	1234212
1234312	1234322			

Lemma 3.9 excludes all possibilities of Φ with more than one element, except the $\{\psi_1$ or ψ_2 or $\psi_3, \psi_7\}$.

(i) $\Phi = \{\psi_1\}$. Then $\mathfrak{n}_\Phi$ is a 27-dimensional commutative algebra, so N_Φ has square integrable representations.

(ii) $\Phi = \{\psi_2\}$. Then $\Delta_\Phi^+ = \{[\psi_2], 2[\psi_2]\}$. We partition $[\psi_2] = S_1 \cup S_2 \cup T_1 \cup T_2$, each part consisting of 8 roots, as follows

S_1:	0110000	0111000	0111100	0111110
	0122101	0122111	0122211	0123211
S_2:	0100000	0111001	0111101	0111111
	0112101	0112111	0112212	0123212
T_1:	1123211	1122211	1122111	1122101
	1111110	1111100	1111000	1110000
T_2:	1123212	1112211	1112111	1112101
	1111111	1111101	1111001	1100000

The sum of 2 roots in $S_1 \cup S_2$ (resp. $T_1 \cup T_2$) is of the form $0\psi_1 + 2\psi_2 + \ldots$ (resp. $2\psi_1 + \ldots$), hence is not a root. If ξ is the j-th root listed in S_1 (resp. in S_2) and η is the j-th root in T_1 (resp. in T_2), then $\xi + \eta = 1233211$ (resp. $= 1223212$), which is in $2[\psi_2]$. No other sum of a root from $S_1 \cup S_2$ with a root from $T_1 \cup T_2$ is equal to 1233211 or 1223212. Choose nonzero linear functionals f_1, f_2 on $\mathfrak{n}_\Phi$ such that $f_1(\mathfrak{g}_\alpha) = 0$ for $\alpha \neq 1233211$ and $f_2(\mathfrak{g}_\beta) = 0$ for $\beta \neq 1223212$. Then $f = f_1 + f_2$ satisfies: b_f is nonsingular on $\mathfrak{n}_{[\psi_2]} \approx \mathfrak{n}_\Phi / \mathfrak{z}_\Phi$, so N_Φ has square integrable representations.

(iii) $\Phi = \{\psi_3\}$. Then $\Delta_\Phi^+ = \{[\psi_3], 2[\psi_3], 3[\psi_3]\}$. $[\psi_3]$ consists of 30 roots and $2[\psi_3]$ consists of 15, so $\dim \mathfrak{n}_{[\psi_3]} \neq \dim \mathfrak{n}_{2[\psi_3]}$, and Lemma 3.8 says that N_Φ cannot have square integrable representations.

(iv) $\Phi = \{\psi_4\}$. Then $\Delta_\Phi^+ = \{[\psi_4], 2[\psi_4], 3[\psi_4], 4[\psi_4]\}$. $[\psi_4]$ consists of 24 roots and $3[\psi_4]$ consists of 8, so $\dim \mathfrak{n}_{[\psi_4]} \neq \dim \mathfrak{n}_{3[\psi_4]}$, and Lemma 3.8 says that N_Φ cannot have square integrable representations.

(v) $\Phi = \{\psi_5\}$. Then $\Delta_\Phi^+ = \{[\psi_5], 2[\psi_5], 3[\psi_5]\}$. $[\psi_5]$ consists of 30 roots and $2[\psi_5]$ consists of 15, so $\dim \mathfrak{n}_{[\psi_5]} \neq \dim \mathfrak{n}_{2[\psi_5]}$, and Lemma 3.8 says that N_Φ cannot have square integrable representations.

(vi) $\Phi = \{\psi_6\}$. Then $\Delta_\Phi^+ = \{[\psi_6], 2[\psi_6]\}$. $2[\psi_6]$ consists of the single root 1234322, and $[\psi_6]$ consists of 16 pairs of roots with sum 1234322:

0000010	and	1234312,	0000110	and	1234212,
0001110	and	1233212,	0001111	and	1233211,
0011110	and	1223212,	0111110	and	1123212,
0011111	and	1223211,	0111111	and	1123211
0012111	and	1222211,	0012211	and	1222111,
0112111	and	1122211,	0112211	and	1122111,
0122111	and	1112211,	0122211	and	1112111,
0123211	and	1111111,	0123212	and	1111110.

Now N_Φ is a 33-dimensional Heisenberg group, so it has square integrable representations.

(vii) $\Phi = \{\psi_7\}$. Then $\Delta_\Phi^+ = \{[\psi_7], 2[\psi_7]\}$, and $[\psi_7]$ consist of 35 roots. Thus $\dim \mathfrak{n}_\Phi/\mathfrak{z}_\Phi$ is odd, so Lemma 3.8 says that N_Φ cannot have square integrable representations.

(viii) $\Phi = \{\psi_1, \psi_7\}$. Then $\Delta^+_\Phi = \{[\psi_1], [\psi_7], [\psi_1] + [\psi_7], 2[\psi_7], [\psi_1] + 2[\psi_7]\}$. Here $[\psi_1]$ consists of the 6 roots 1000000, 1100000, 1110000, 1111000, 1111100 and 1111110, while $2[\psi_7]$ consists just of the root 0123212. Thus $\dim \mathfrak{n}_{[\psi_1]} \neq \dim \mathfrak{n}_{2[\psi_7]}$, and Lemma 3.8 says that N_Φ cannot have square integrable representations.

(ix) $\Phi = \{\psi_2, \psi_7\}$. Then $\Delta^+_\Phi = \{[\psi_2], [\psi_7], [\psi_2] + [\psi_7], [\psi_2] + 2[\psi_7], 2[\psi_2] + [\psi_7], 2[\psi_2] + 2[\psi_7]\}$. Here $[\psi_2]$ consists of the 10 roots 0100000, 1100000, 0110000, 111000, 0111000, 1111000, 0111100, 1111100, 0111110 and 1111110, while $[\psi_2] + 2[\psi_7]$ consists only of 0123212 and 1123212. So $\dim \mathfrak{n}_{[\psi_2]} \neq \dim \mathfrak{n}_{[\psi_2]+2[\psi_7]}$, and Lemma 3.8 says that N_Φ cannot have square integrable representations.

(x) $\Phi = \{\psi_3, \psi_7\}$. Then $\Delta^+_\Phi = \{[\psi_3], [\psi_7], [\psi_3] + [\psi_7], 2[\psi_3] + [\psi_7], 3[\psi_3] + [\psi_7], 2[\psi_3] + 2[\psi_7], 3[\psi_3] + 2[\psi_7]\}$. Here $[\psi_3]$ consists of 12 roots while $2[\psi_3] + 2[\psi_7]$ consists of 3, so $\dim \mathfrak{n}_{[\psi_3]} \neq \dim \mathfrak{n}_{2[\psi_3]+2[\psi_7]}$, and Lemma 3.8 says that N_Φ cannot have square integrable representations.

6.5. Split E_8: 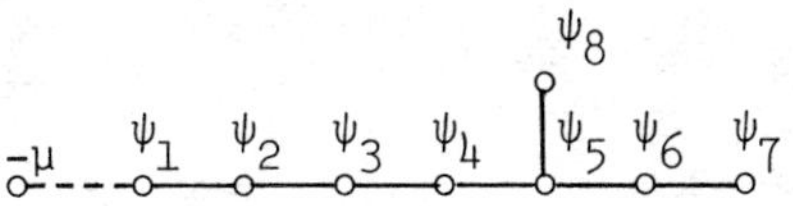

. Write ghijkℓmn for a root $g\psi_1 + h\psi_2 + i\psi_3 + j\psi_4 + k\psi_5 + \ell\psi_6 + m\psi_7 + n\psi_8$. Then the positive roots are (i) the ones listed for E_6 in §6.3, each preceded by two zeroes, (ii) the additional ones listed for E_7 in §6.4, each preceded by a zero, and (iii) the roots labeled

10000000	11000000	11100000	11110000	11111000
11111100	11111001	11111110	11111101	11111111
11112101	11112111	11122101	11112211	11122111
11222101	11122211	11222111	12222101	11123211
11222211	12222111	11123212	11223211	12222211
11223212	11233211	12223211	11233212	12223212
12233211	11234212	12233212	12333211	11234312
12234212	12333212	11234322	12234312	12334212
12234322	12334312	12344212	12334322	12344312
12345312	12344322	12345313	12345322	12345323
12345422	12345423	12346423	12356423	12456423
13456423	23456423			

Lemma 3.9 excludes all possibilities of Φ with more than one element, except the $\{\psi_6$ or $\psi_7, \psi_8\}$.

(i) $\Phi = \{\psi_1\}$. Then $\Delta_\Phi^+ = \{[\psi_1], 2[\psi_1]\}$, $2[\psi_1]$ consists of a single root 23456423, and $[\psi_1]$ consists of 56 roots that fall into 28 pairs with sum 23456423. So N_Φ is the 57-dimensional Heisenberg group, and thus has square integrable representations.

(ii) $\Phi = \{\psi_2\}$. Then $\Delta_\Phi^+ = \{[\psi_2], 2[\psi_2], 3[\psi_2]\}$. $[\psi_2]$ consists of 54 roots and $2[\psi_2]$ consists of 27, so $\dim \mathfrak{n}_{[\psi_2]} \neq \dim \mathfrak{n}_{2[\psi_2]}$ and Lemma 3.8 says that N_Φ cannot have square integrable representations.

(iii) $\Phi = \{\psi_3\}$. Then $\Delta_\Phi^+ = \{[\psi_3], 2[\psi_3], 3[\psi_3], 4[\psi_3]\}$. $[\psi_3]$ consists of 48 roots and $3[\psi_3]$ consists of 16, so $\dim \mathfrak{n}_{[\psi_3]} \neq \dim \mathfrak{n}_{3[\psi_3]}$ and Lemma 3.8 says that N_Φ cannot have square integrable representations.

(iv) $\Phi = \{\psi_4\}$. Then $\Delta_\Phi^+ = \{[\psi_4], 2[\psi_4], 3[\psi_4], 4[\psi_4], 5[\psi_4]\}$. $[\psi_4]$ consists of 40 roots and $4[\psi_4]$ consists of 10, so $\dim \mathfrak{n}_{[\psi_4]} \neq \dim \mathfrak{n}_{4[\psi_4]}$ and Lemma 3.8 says that N_Φ cannot have square integrable representations.

(v) $\Phi = \{\psi_5\}$. Then $\Delta_\Phi^+ = \{[\psi_5], 2[\psi_5], 3[\psi_5], 4[\psi_5], 5[\psi_5], 6[\psi_5]\}$, $2[\psi_5]$ consists of 30 roots and $4[\psi_5]$ consists of 15, so $\dim \mathfrak{n}_{2[\psi_5]} \neq \dim \mathfrak{n}_{4[\psi_5]}$ and Lemma 3.8 says that N_Φ cannot have square integrable representations.

(vi) $\Phi = \{\psi_6\}$. Then $\Delta_\Phi^+ = \{[\psi_6], 2[\psi_6], 3[\psi_6], 4[\psi_6]\}$. $[\psi_6]$ consists of 42 roots and $3[\psi_6]$ consists of 14, so $\dim \mathfrak{n}_{[\psi_6]} \neq \dim \mathfrak{n}_{3[\psi_6]}$ and Lemma 3.8 says that N_Φ cannot have square integrable representations.

(vii) $\Phi = \{\psi_7\}$. Then $\Delta_\Phi^+ = \{[\psi_7], 2[\psi_7]\}$. Here $2[\psi_7]$ consists of 14 roots, including the maximal root 23456423, and $[\psi_7]$ consists of 64 roots that fall into 32 pairs with sum 23456423. If f is a nonzero linear functional on $\mathfrak{n}$ with $f(\mathfrak{g}_\alpha) = 0$ for $\alpha \neq \mu$, then b_f is nonsingular on $\mathfrak{n}_{[\psi_7]} \approx \mathfrak{n}_\Phi/\mathfrak{z}_\Phi$, so N_Φ has square integrable representations.

(viii) $\Phi = \{\psi_8\}$. Then $\Delta_\Phi^+ = \{[\psi_8], 2[\psi_8], 3[\psi_8]\}$. $[\psi_8]$ consists of 56 roots and $2[\psi_8]$ consists of 28, so $\dim \mathfrak{n}_{[\psi_8]} \neq \dim \mathfrak{n}_{2[\psi_8]}$, and Lemma 3.8 says that N_Φ cannot have square integrable representations.

(ix) $\Phi = \{\psi_6, \psi_8\}$. The highest $\mathfrak{a}_\Phi$-root is $4[\psi_6] + 3[\psi_8]$. $[\psi_8]$ consists of the 6 roots 00000001, 00001001, 00011001, 00111001, 01111001 and 11111001, while $4[\psi_6] + 2[\psi_8]$ consists just of 12345422. So $\dim \mathfrak{n}_{[\psi_8]} \neq \dim \mathfrak{n}_{4[\psi_6]+2[\psi_8]}$, and Lemma 3.8 says that N_Φ cannot have square integrable representations.

(x) $\Phi = \{\psi_7, \psi_8\}$. The highest $\mathfrak{a}_\Phi$-root is $2[\psi_7] + 3[\psi_8]$. $[\psi_7]$ consist of the 7 roots 00000010, 00000110, 00001110, 00011110, 00111110, 01111110 and 11111110, and $[\psi_7] + 3[\psi_8]$ consists just of 12345313. So $\dim \mathfrak{n}_{[\psi_7]} \neq \dim \mathfrak{n}_{[\psi_7]+3[\psi_8]}$, and Lemma 3.8 says that N_Φ cannot have square integrable representations.

6.6. Summary-exceptional split case. We now summarize the results of §6 using [29, pp. 282-282b] for the representations of $\mathfrak{m}_{\Phi'}$.

Our notation for the exceptional groups: the second subscript denotes the Cartan classification type of the maximal compact subgroup. Thus G_{2,A_1A_1} is a split G_2, F_{4,B_4} is an F_4 of real rank 1, E_{7,E_6T_1} is the automorphism group of the 27-dimensional bounded symmetric domain, etc.

Here are the parabolic subgroups -- or, rather, their identity components -- of the real split simple exceptional Lie groups, in which the nilradical has square integrable representations.

6.6.1. $G = G_{2,A_1A_1}$: $\overset{\psi_1}{\circ}\!\Rrightarrow\!\overset{\psi_2}{\circ}$, real group of type G_2 with maximal compact subgroup $K \cong SO(4)$, automorphism group of the split real Cayley algebra $\mathfrak{C}_{\mathbb{R}}$. Here G acts faithfully on $\operatorname{Im} \mathfrak{C}_{\mathbb{R}} \cong \mathbb{R}^{3,4}$ by $\circ\!\Rrightarrow\!\overset{1}{\circ}$.

(i) $\Phi = \{\psi_1\}$. Then the identity component $P_\Phi^0 = N_\Phi A_\Phi M_{\Phi'}^0$, where N_Φ is the 5-dimensional Heisenberg group, $M_{\Phi'}^0 \cong SL(2;\mathbb{R})$, acting on trivially on $\mathfrak{z}_\Phi$ and acting on $\mathfrak{n}_\Phi/\mathfrak{z}_\Phi$ by $\overset{3}{\circ}$.

6.6.2. $G = F_{4,C_3A_1}$: 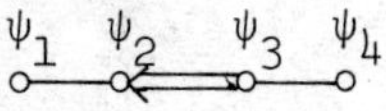 , real group of type F_4 with maximal compact subgroup $K \cong Sp(3)\cdot Sp(1)$.

(i) $\Phi = \{\psi_1\}$. Then $P_\Phi^0 = N_\Phi A_\Phi M_{\Phi'}^0$ where $N_\Phi \cong \operatorname{Im}\mathfrak{C}_{\mathbb{R}} + \mathfrak{C}_{\mathbb{R}}$ with $(z,x)(z',x') = (z + z' + \operatorname{Im} x\bar{x}', x+x')$, where $M_{\Phi'}^0 \cong Spin(3,4)^0$, and where $M_{\Phi'}^0$ acts on $\mathfrak{C}_{\mathbb{R}}$ by the spin representation 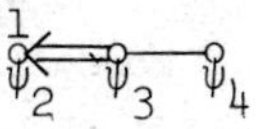 and on $\operatorname{Im}\mathfrak{C}_{\mathbb{R}}$ by the vector representation 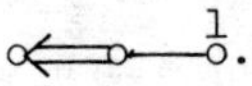. The fact that this is an action by automorphisms on N_Φ follows from the Clifford algebra construction of Spin groups.

Here is a model of P_Φ^0 more in accordance with phenomena that occur in the E series. Let σ denote the spin representation of $Spin(3,4)^0$ and let ν denote the vector representation. Then σ is real, say a representation on $\mathbb{R}^8$, and ν has multiplicity 1 in $\Lambda^2(\sigma)$, say on $\mathbb{R}^{3,4} \subset \Lambda^2(\mathbb{R}^8)$. Let $p\colon \Lambda^2(\mathbb{R}^8) \to \mathbb{R}^{3,4}$ be the equivariant projection. Then

$$P_\Phi^0 \cong (\mathbb{R}^{3,4} + \mathbb{R}^8)\cdot(Spin(3,4)^0 \times \mathbb{R}^+)$$

with composition $(z,x,g,a)(z',x',g',a') = (z+a^2\nu(g)z' + a\cdot p(x \wedge \sigma(g)x'), x + a\cdot\sigma(g)x', gg', aa')$.

(ii) $\Phi = \{\psi_4\}$. Then $P_\Phi^0 = N_\Phi A_\Phi M_{\Phi'}^0$ where $N_\Phi \cong \mathbb{R} + \mathbb{R}^{14}$ is the 15-dimensional Heisenberg group, $M_{\Phi'}^0 \cong Sp(3;\mathbb{R})$, and $M_{\Phi'}^0$ acts on $\mathbb{R}^{14}$ by the subrepresentation 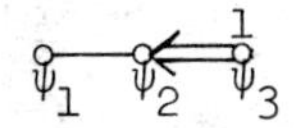 of $\Lambda^3($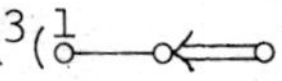$)$ =  $\oplus$ 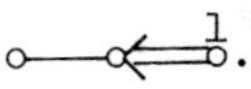. That subrepresentation has an antisymmetric bilinear invariant, which specifies the composition in N_Φ .

6.6.3. $G = E_{6,C_4}$: 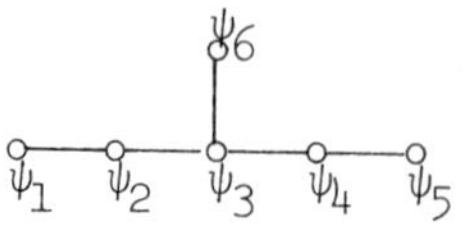 , real group of type E_6 with maximal compact subgroup $K \cong Sp(4)/\{\pm I\}$.

(i) $\Phi = \{\psi_1\}$ and $\Phi = \{\psi_5\}$. Then $P_\Phi^0 = N_\Phi A_\Phi M_{\Phi'}^0$, where $N_\Phi \cong \mathbb{R}^{16}$, commutative, where $M_{\Phi'}^0 \cong Spin(5,5)^0$, and where $M_{\Phi'}^0$ acts by the half spin representation 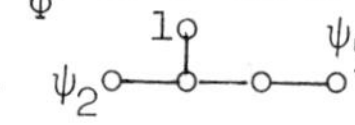 when $\Phi = \{\psi_1\}$, 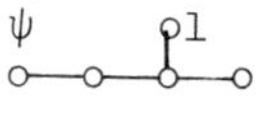 when $\Phi = \{\psi_5\}$.

(ii) $\Phi = \{\psi_6\}$. Then $P_\Phi^0 = N_\Phi A_\Phi M_{\Phi'}^0$, where $N_\Phi \cong \mathbb{R} + \mathbb{R}^{20}$ is the 21-dimensional Heisenberg group, $M_{\Phi'}^0 \cong SL(6;\mathbb{R})$, and $M_{\Phi'}^0$ acts on $\mathbb{R}^{20}$ by 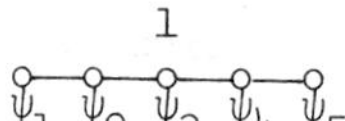 , which is Λ^3 of the vector representation 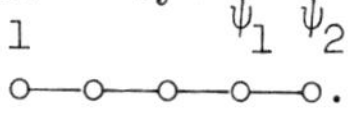.

(iii) $\Phi = \{\psi_1, \psi_5\}$. Then $P_\Phi^0 = N_\Phi A_\Phi M_{\Phi'}^0$, where $N_\Phi \cong \left\{\begin{pmatrix} 1 & x & z \\ 0 & 1 & y \\ 0 & 0 & 1 \end{pmatrix} : x, y, z \in \mathfrak{C}_{\mathbb{R}}\right\}$ and $M_{\Phi'}^0 \cong Spin(4,4)^0$. Here x describes $\mathfrak{n}_{[\psi_1]}$, y describes $\mathfrak{n}_{[\psi_5]}$ and z describes the center $\mathfrak{n}_{[\psi_1]+[\psi_5]}$, and $M_{\Phi'}^0$ acts by

$$Ad(g): \begin{pmatrix} 1 & x & z \\ 0 & 1 & y \\ 0 & 0 & 1 \end{pmatrix} = \begin{pmatrix} 1 & \sigma_+(g)x & \nu(g)z \\ 0 & 1 & \sigma_-(g)y \\ 0 & 0 & 1 \end{pmatrix}$$

where σ_+: 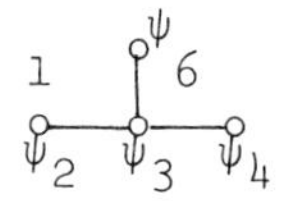 and σ_-: 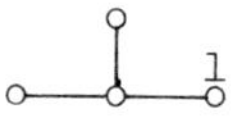 are the half spin representations and ν: 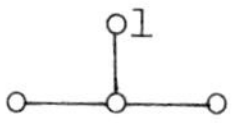 is the vector representation. The fact that this is an action by automorphisms on N_Φ is equivalent to the Triality Principle.

One can describe P_Φ^0 without Cayley numbers, using the fact that 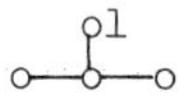 is a subrepresentation of multiplicity 1 in 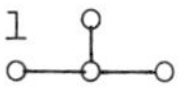 $\otimes$ 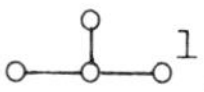, just as in (6.6.2)(i): if $p: \mathbb{R}^{4,4} \otimes \mathbb{R}^{4,4} \to \mathbb{R}^{4,4}$ is the equivariant projection then

$$P_\Phi^0 \cong (\mathbb{R}^{4,4} + \mathbb{R}^{4,4} + \mathbb{R}^{4,4})\cdot(SO(4,4)^0 \times \mathbb{R}^+)$$

with composition $(z,y,x,g,a)(z',y',x',g',a')$

$= (z + a^2\nu(g)z' + a\cdot p(x \otimes \sigma_-(g)y'),\ y + a\cdot\sigma_-(g)y',\ x + a\cdot\sigma_+(g)x',\ gg', aa')$.

<u>6.6.4.</u> $G = E_{7,A_7}$: 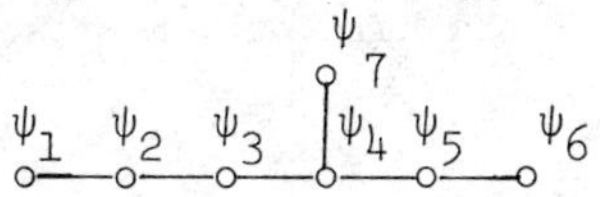 , real group of type E_7 with maximal compact subgroup $K \cong SU(8)/\{\pm I\}$.

(i) $\Phi = \{\psi_1\}$. Then $P_\Phi^0 = N_\Phi A_\Phi M_{\Phi'}^0$ where $N_\Phi \cong \mathbb{R}^{27}$, commutative, where $M_{\Phi'}^0 \cong E_{6,C_4}$, and where $M_{\Phi'}^0$ acts on N_Φ by

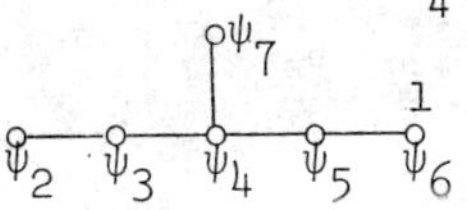 .

(ii) $\Phi = \{\psi_2\}$. Then $P_\Phi^0 = N_\Phi A_\Phi M_{\Phi'}^0$ where $N_\Phi \cong \mathbb{R}^{5,5} + (\mathbb{R}^2 \otimes \mathbb{R}^{16})$ and $M_{\Phi'}^0 \cong SL(2;\mathbb{R}) \times Spin(5,5)^0$ as follows. $M_{\Phi'}^0$ acts on $\mathbb{R}^2 \otimes \mathbb{R}^{16}$ by $\alpha \otimes \sigma_+$: 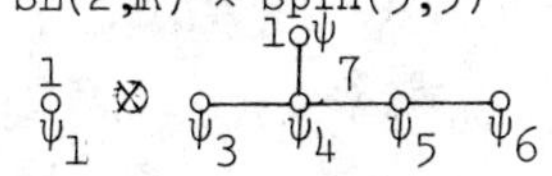 , on $\mathbb{R}^{5,5}$ by $1 \otimes \nu$: 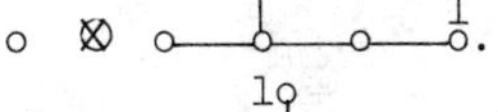.
The latter is a subrepresentation of multiplicity 1 in $\Lambda^2($ 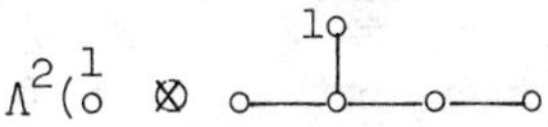 $)$; let $p: \Lambda^2(\mathbb{R}^2 \otimes \mathbb{R}^{16}) \to \mathbb{R}^{5,5}$ denote the equivariant projection. Now

$$P_\Phi^0 \cong \{\mathbb{R}^{5,5} + (\mathbb{R}^2 \otimes \mathbb{R}^{16})\}\cdot\{(SL(2;\mathbb{R}) \times Spin(5,5)^0) \times \mathbb{R}^+\}$$

with composition $(z,x,\gamma,g,a)(z',x',\gamma',g',a') =$

$(z + a^2\nu(g)z' + a\cdot p(x \wedge [\alpha(\gamma) \otimes \sigma_+(g)]x',\ x + a[\alpha(\gamma) \otimes \sigma_+(g)]x',$

$\gamma\gamma', gg', aa')$.

(iii) $\Phi = \{\psi_6\}$. Then $P_\Phi^0 = N_\Phi A_\Phi M_{\Phi'}^0$ where $N_\Phi \cong \mathbb{R} + \mathbb{R}^{32}$ is the 33-dimensional Heisenberg group, $M_{\Phi'}^0 \cong Spin(6,6)^0$, $M_{\Phi'}^0$ acts on $\mathbb{R}^{32}$ by the half spin representation

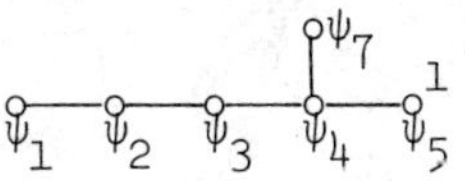 . That half spin representation has an antisymmetric

bilinear invariant, which specifies the composition in N_Φ.

6.6.5. $G = E_{8,D_8}$: 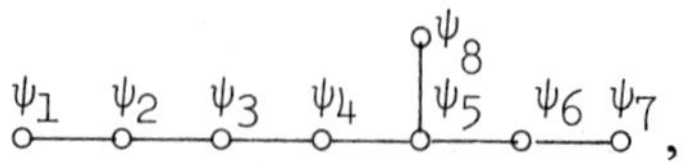

, real group of type E_8 with maximal compact subgroup $K \cong Spin(16)/\{1,e\}$, where $\{1, e\}$ is the kernel of the half spin representation

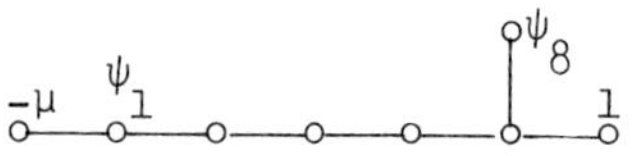

.

(i) $\Phi = \{\psi_1\}$. Then $P_\Phi^0 = N_\Phi A_\Phi M_{\Phi'}^0$, where $N_\Phi \cong \mathbb{R} + \mathbb{R}^{56}$ is the 57-dimensional Heisenberg group, $M_{\Phi'}^0 \cong E_{7,A_7}$, and $M_{\Phi'}^0$ acts on $\mathbb{R}^{56}$ by

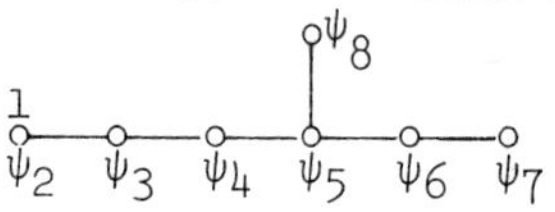

. That representation has an antisymmetric bilinear invariant, which specifies the composition in N_Φ.

(ii) $\Phi = \{\psi_7\}$ Then $P_\Phi^0 = N_\Phi A_\Phi M_{\Phi'}^0$, where $N_\Phi \cong \mathbb{R}^{7,7} + \mathbb{R}^{64}$ and $M_{\Phi'}^0 \cong Spin(7,7)^0$ as follows. $M_{\Phi'}^0$ acts on $\mathbb{R}^{64}$ by the half spin representation σ_+: 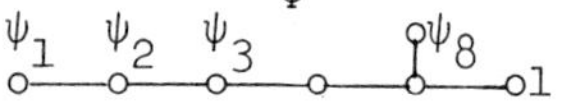

on $\mathbb{R}^{7,7}$ by the vector representation ν: 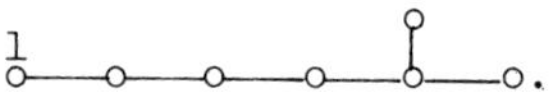

. Here ν is a subrepresentation of multiplicity 1 in $\Lambda^2(\sigma_+)$; let $p: \Lambda^2(\mathbb{R}^{64}) \to \mathbb{R}^{7,7}$ be the equivariant projection. Then

$$P_\Phi^0 \cong (\mathbb{R}^{7,7} + \mathbb{R}^{64})\cdot(Spin(7,7)^0 \times \mathbb{R}^+)$$

with composition $(z,x,g,a)(z',x',g',a') = (z + a^2\nu(g)z' + a\cdot p(x \wedge \sigma_+(g)x'),$ $x + a\cdot\sigma_+(g)x', gg', aa')$.

§7. Passage to the General Exceptional Group

We now carry the classification of §6 over to arbitrary exceptional real simple Lie groups.

7.1. Type G_2. The complex case is the only noncompact non-split case. There we just have $\Phi = \{\beta_1\}$ in the simple $\mathfrak{a}$-root system

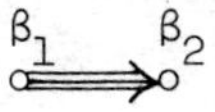

, and

$$P_\Phi \cong (\mathbb{C} + \mathbb{C}^4)\cdot GL(2;\mathbb{C})$$

with composition $(z,x,g)(z',x',g') =$
$(z + (\det g)^3 z' + \mathfrak{a}(x,S^3(g)x'),\ x + S^3(g)x', gg')$ where $S^3(g)$ denotes the action of $g \in GL(2;\mathbb{C})$ on the $\mathbb{C}^4$ of degree 3 polynomials on $\mathbb{C}^2$ and $\mathfrak{a}$ is the antisymmetric bilinear invariant of the representation S^3.

7.2. Type F_4. Here are the noncompact non-split cases.

7.2.1. $G = (F_4)_{\mathbb{C}}$, complex form. Then (6.6.2)(i) gives the case

$$P_\Phi = (\operatorname{Im}\mathfrak{C}_{\mathbb{C}} + \mathfrak{C}_{\mathbb{C}})\cdot(\mathrm{Spin}(7,\mathbb{C}) \times \mathbb{C}^*)$$

where $\mathfrak{C}_{\mathbb{C}}$ is the complex Cayley algebra and $\operatorname{Im}\mathfrak{C}_{\mathbb{C}}$ is the (-1)-eigenspace for conjugation $x \mapsto \bar{x}$ of $\mathfrak{C}_{\mathbb{C}}$ over $\mathbb{C}$, and where $\mathbb{C}^*$ is the multiplicative group of nonzero complex numbers. The composition in P_Φ is $(z,x,g,\gamma)(z',x',g',\gamma') =$ $(z + \gamma^2\nu(g)z' + \gamma\ \mathrm{Im}\ x\cdot\overline{\sigma(g)x'},\ x + \gamma\sigma(g)x', gg', \gamma\gamma')$ where $\nu\colon \mathrm{Spin}(7;\mathbb{C}) \to SO(7;\mathbb{C})$ is the vector representation and σ is the spin representation. And (6.6.2)(ii) gives the case

$$P_\Phi = (\mathbb{C} + \mathbb{C}^{14})\cdot(Sp(3;\mathbb{C}) \times \mathbb{C}^*),$$

There, the composition is $(z,x,g,\gamma)(z',x',g',\gamma') =$ $(z + \gamma^2 z' + \gamma\cdot\mathfrak{A}(x,\tau(g)x'),\ x + \gamma\cdot\tau(g)x', gg', \gamma\gamma')$ where τ is the representation 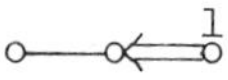 of $Sp(3;\mathbb{C})$ on $\mathbb{C}^{14}$ and $\mathfrak{A}$ is its antisymmetric bilinear invariant.

7.2.2. $G = F_{4,B_4}$, real group of type F_4 and real rank 1. Its maximal compact subgroup $K \cong \mathrm{Spin}(9)$ and the Satake diagram is 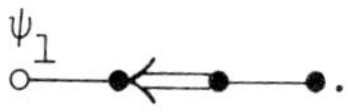. So (6.6.2) gives only the case $\Phi = \{\beta_1\}$, $\beta_1 = \psi_1|_{\mathfrak{a}}$. There

$$P_\Phi \cong (\mathrm{Im}\,\mathfrak{C} + \mathfrak{C})\cdot(\mathrm{Spin}(7) \times \mathbb{R}^*)$$

where $\mathfrak{C}$ is the Cayley division algebra over $\mathbb{R}$. The composition is as in (7.2.1).

7.3. Type E_6. Here are the noncompact non-split cases.

7.3.1. $G = (E_6)_{\mathbb{C}}$, complex form. Then (6.6.3)(i) gives the case

$$P_\Phi \cong \mathbb{C}^{16}\cdot(\mathrm{Spin}(10;\mathbb{C}) \times \mathbb{C}^*)$$

with composition $(z,g,\gamma)(z',g',\gamma') = (z + \gamma\cdot\sigma_+(g)z', gg', \gamma\gamma')$ where σ_+ is a half spin representation of $\mathrm{Spin}(10;\mathbb{C})$ on $\mathbb{C}^{16}$. Also, (6.6.3)(ii) gives the case

$$P_\Phi = (\mathbb{C} + \mathbb{C}^{20})\cdot(SL(6;\mathbb{C}) \times \mathbb{C}^*/\mathbb{Z}_6)$$

with composition $(z,x,g,\gamma)(z',x',g',\gamma') = (z + \gamma^2 z' + \gamma\cdot\mathcal{A}(x,\lambda(g)x'), x + \gamma\cdot\lambda(g)x', gg', \gamma\gamma')$, where λ is the representation 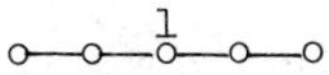

of $SL(6;\mathbb{C})$ on $\mathbb{C}^{20}$ and $\mathcal{A}$ is its antisymmetric bilinear invariant. Finally, (6.6.3)(iii) gives the case

$$P_\Phi \cong \left\{\begin{pmatrix} 1 & x & z \\ 0 & 1 & y \\ 0 & 0 & 1 \end{pmatrix} : x, y, z \in \mathcal{C}_{\mathbb{C}}\right\}\cdot(\mathrm{Spin}(8;\mathbb{C}) \times \mathbb{C}^*)$$

with composition $(z,y,x,g,\gamma)(z',y',x',g',\gamma') =$
$(z + \gamma^2\cdot\nu(g)z' + \gamma x\cdot\sigma_-(g)y', y + \gamma\cdot\sigma_-(g)y', x + \gamma\cdot\sigma_+(g)x', gg', \gamma\gamma')$.

<u>7.3.2.</u> $G = E_{6,A_1A_5}$, real group of type E_6 and real rank 4. Its maximal compact subgroup $K \cong \{SU(2) \times SU(6)\}/\{\pm(I_2,I_6), \pm(I_2,e^{2\pi i/3}I_6), \pm(I_2,e^{-2\pi i/3}I_6)\}$, and its Satake diagram is

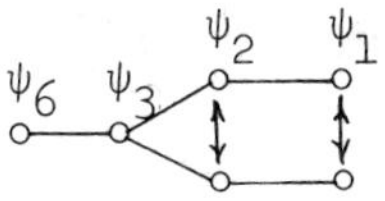

. Here (6.6.3)(i) does not define a parabolic subgroup. But (6.6.3)(ii) gives the case

$$P_\Phi \cong (\mathbb{R} + \mathbb{R}^{20})\cdot\{(SU(3,3)/\mathbb{Z}_3) \times \mathbb{R}^+\}$$

with composition as described above. And (6.6.3)(iii) gives the case

$$P_\Phi \cong \left\{\begin{pmatrix} 1 & x & z \\ 0 & 1 & y \\ 0 & 0 & 1 \end{pmatrix} : x, y, z \in \mathfrak{C}_{\mathbb{R}}\right\}\cdot(\mathrm{Spin}(3,5) \times \mathbb{R}^+)$$

with composition as described above.

7.3.3. $G = E_{6,D_5T_1}$, real group of type E_6 with maximal compact subgroup $K \cong \{\mathrm{Spin}(10) \times SO(2)\}/\mathbb{Z}_2$ where $\mathbb{Z}_2 = \{(1,1), (-1,-1)\}$. The Satake diagram is

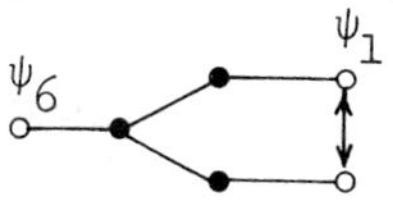

. Here (6.6.3)(i) does not define a parabolic subgroup of G. But (6.6.3)(ii) gives the case

$$P_\Phi \cong (\mathbb{R} + \mathbb{R}^{20})\cdot\{(SU(1,5)/\mathbb{Z}_3) \times \mathbb{R}^+\}$$

with composition as described above. And (6.6.3)(iii) gives the case

$$P_\Phi \cong \left\{\begin{pmatrix} 1 & x & z \\ 0 & 1 & y \\ 0 & 0 & 1 \end{pmatrix} : x, y\ z \in \mathfrak{C}\right\}\cdot(\mathrm{Spin}\ (1,7) \times \mathbb{R}^+)$$

with composition described above.

7.3.4. $G = E_{6,F_4}$, real group of type E_6 with maximal compact subgroup $K \cong F_4$. The Satake diagram is

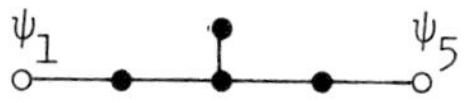

. Here (6.6.3)(i) gives the case

$$P_\Phi \cong \mathbb{R}^{16}\cdot\{\mathrm{Spin}(1,9) \times \mathbb{R}^+\}$$

with composition as above. The case (6.6.3)(ii) does not define a parabolic subgroup of G. And (6.6.3)(iii) gives the case

$$P_\Phi \cong \left\{\begin{pmatrix} 1 & x & z \\ 0 & 1 & y \\ 0 & 0 & 1 \end{pmatrix} : x, y, z \in \mathfrak{C}\right\} \cdot (\mathrm{Spin}(8) \times \mathbb{R}^+)$$

with composition as above.

7.4. Type E_7. Here are the noncompact non-split cases.

7.4.1 $G = (E_7)_{\mathbb{C}}$, complex form. Then (6.6.4)(i) gives the case

$$P_\Phi = \mathbb{C}^{27} \cdot \{(E_6)_{\mathbb{C}} \times \mathbb{C}^*\}$$

with composition $(z,g,\gamma)(z',g',\gamma') = (z+\gamma\cdot\nu(g)z', gg', \gamma\gamma')$ where ν is the representation 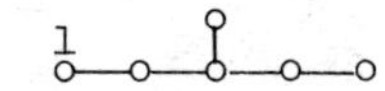 of $(E_6)_{\mathbb{C}}$. From (6.6.4)(ii) we have the case

$$P_\Phi = \{\mathbb{C}^{10} + (\mathbb{C}^2 \otimes \mathbb{C}^{16})\} \cdot \{SL(2;\mathbb{C}) \times \mathrm{Spin}(10;\mathbb{C}) \times \mathbb{C}^*\}$$

with composition $(z,x;\gamma,g,a)(z',x';\gamma',g',a') =$ $(z + a^2\nu(g)z' + ap(x \wedge [\alpha(\gamma) \otimes \sigma_+(g)]x', x + a[\alpha(\gamma) \otimes \sigma_+(g)]x', \gamma\gamma', gg', aa')$ where ν represents $\mathrm{Spin}(10;\mathbb{C})$ by 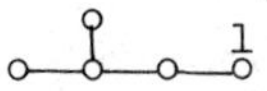, $\alpha \otimes \sigma_+$ represents $SL(2;\mathbb{C}) \times \mathrm{Spin}(10;\mathbb{C})$ by  $\otimes$ 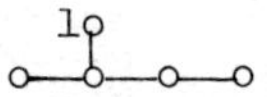, and p is projection of $\Lambda^2(\mathbb{C}^2 \otimes \mathbb{C}^{10})$ onto the unique subspace for the subrepresentation $1 \otimes \nu$ of $\Lambda^2(\alpha \otimes \sigma_+)$. Finally, (6.6.4)(iii) gives us the case

$$P_\Phi \cong (\mathbb{C} + \mathbb{C}^{32})\cdot(\mathrm{Spin}(12;\mathbb{C}) \times \mathbb{C}^*)$$

with composition $(z,x,g,\gamma)(z',x',g',\gamma')$

$= (z + \gamma^2 z' + \gamma\mathfrak{a}(x,\sigma_+(g)x'),\ x + \gamma\sigma_+(g)x', gg', \gamma\gamma')$ where

σ_+: 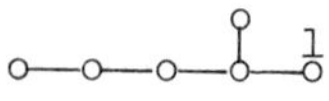

and $\mathfrak{a}$ is its antisymmetric bilinear invariant.

<u>7.4.2.</u> $G = E_{7,D_6A_1}$, real group of type E_7 with maximal compact subgroup $K \cong SO(12) \times SU(2)$. The Satake diagram is

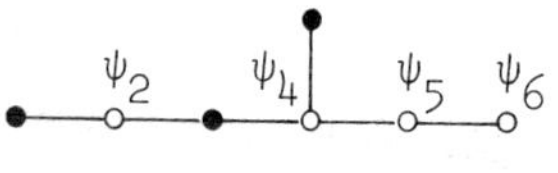

. Here (6.6.4)(i) does not define a parabolic subgroup of G. (6.6.4)(ii) defines the parabolic

$$P_\Phi \cong \{\mathbb{R}^{3,7} + (\mathbb{R}^2 \otimes \mathbb{R}^{16})\}\cdot\{SU(2) \times \mathrm{Spin}(3,7) \times \mathbb{R}^+\}$$

with composition as above. And (6.6.4)(iii) gives the case

$$P_\Phi \cong \{\mathbb{R} + \mathbb{R}^{32}\}\cdot\{\mathrm{Spin}^*(12) \times \mathbb{R}^+\}$$

where $SO^*(2k)$ is the real form of $SO(2k;\mathbb{C})$ with maximal compact subgroup $U(k)$, and $\mathrm{Spin}^*(2k)$ denotes its inverse image in $\mathrm{Spin}(2k;\mathbb{C})$. Composition as above.

<u>7.4.3.</u> $G = E_{7,E_6T_1}$, real group of type E_7 with maximal compact subgroup $K \cong E_6 \times SO(2)$. The Satake diagram is

. Here (6.6.4)(i) gives the case

$$P_\Phi \cong \mathbb{R}^{27} \cdot \{E_{6,F_4} \times \mathbb{R}^+\}$$

with composition $(z,g,\gamma)(z',g',\gamma') = (z + \gamma \cdot \nu(g)z', gg', \gamma\gamma')$ with

ν: 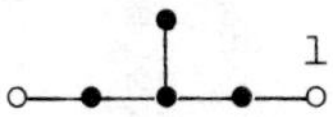 as above. (6.6.4)(ii) gives the case

$$P_\Phi \cong \{\mathbb{R}^{1,9} + (\mathbb{R}^2 \otimes \mathbb{R}^{16})\} \cdot \{SL(2;\mathbb{R}) \times Spin(1,9) \times \mathbb{R}^+\}$$

with composition described above. And (6.6.4)(iii) gives

$$P_\Phi \cong \{\mathbb{R} + \mathbb{R}^{32}\} \cdot \{Spin(2, 10) \times \mathbb{R}^+\}$$

where again the composition is specified above.

7.5. Type E_8. Here are the noncompact non-split cases.

7.5.1. $G = (E_8)_{\mathbb{C}}$, complex form. Then (6.6.5)(i) gives the case

$$P_\Phi \cong (\mathbb{C} + \mathbb{C}^{56}) \cdot (\{(E_7)_{\mathbb{C}} \times \mathbb{C}^*\}/\{(1,1), (-1, -1)\})$$

where the group law is $(z,x,g,\gamma)(z',x',g',\gamma') =$
$(z + \gamma^2 z' + \gamma \cdot \mathcal{A}(x, \nu(g)x'), x + \gamma \cdot \nu(g)x', gg', \gamma\gamma')$. Here ν is the representation 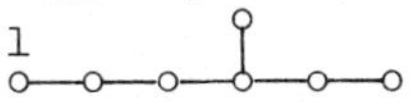 of $(E_7)_{\mathbb{C}}$ on $\mathbb{C}^{56}$ and $\mathcal{A}$ is its antisymmetric bilinear invariant. Also, (6.6.5)(ii) gives the case

$$P_\Phi \cong (\mathbb{C}^{14} + \mathbb{C}^{64}) \cdot (\{Spin\,(14;\mathbb{C}) \times \mathbb{C}^*\}/\{(1,1),(-1,-1)\})$$

with composition $(z,x,g,\gamma)(z',x',g',\gamma') =$

$(z + \gamma^2\nu(g)z' + \gamma\cdot p(x \wedge \sigma_+(g)x'),\ x + \gamma\cdot\sigma_+(g)x', gg', \gamma\gamma')$ where σ_+ is the half spin representation 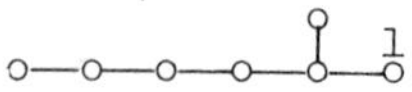

of Spin $(14;\mathbb{C})$ on $\mathbb{C}^{64}$, ν is the vector representation 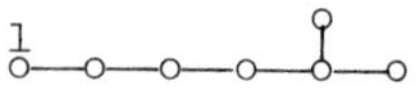

, and $p: \Lambda^2(\mathbb{C}^{64}) \to \mathbb{C}^{14}$ is projection to the ν-isotypic subspace.

7.5.2. $G = E_{8,E_7A_1}$, real group of type E_8 with maximal compact subgroup $\{E_7 \times SU(2)\}/\{(1,1), (-1,-1)\}$. The Satake diagram is 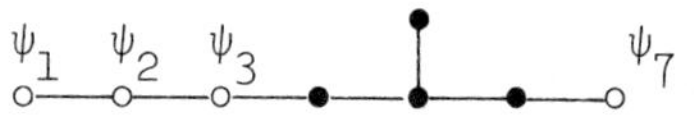

. Then (6.6.5)(i) gives us

$$P_\Phi \cong (\mathbb{R} + \mathbb{R}^{56})\cdot(E_{7,E_6T_1} \times \mathbb{R}^+)$$

with composition as above, and (6.6.5)(ii) gives the case

$$P_\Phi \cong (\mathbb{R}^{3,11} + \mathbb{R}^{64})\cdot(\mathrm{Spin}\ (3,11) \times \mathbb{R}^+)$$

again with composition described above.

§8. Three Consequences of the Classification

8.1. Two-step nilpotency. A glance at the results in §§4-7 gives

Theorem. Let G be a real or complex reductive Lie group, let N be the nilradical of a parabolic subgroup, and suppose that N has square integrable representations. Then N is abelian or 2-step nilpotent.

Evidently, a priori knowledge of this result would have simplified the classification.

8.2. Boundary components. Another run through the classification, together with [24 , Theorem 4.13], gives

Theorem. Consider a bounded symmetric domain $D = G/K$. If F is a boundary component of D, then the nilradical of the maximal parabolic subgroup

$$P_F = \{g \in G: gF = F\}$$

has square integrable representations.

If F is a point on the Bergman-Shilov boundary of D, and if we decompose $P_F = M_F A_F N_F$, then $\dim(M_F A_F) \geqslant \dim K$, with equality if and only if D is of tube type. In the latter case,

$(\mathfrak{m}_F + \mathfrak{a}_F)_{\mathbb{C}}$ *is related to* $\mathfrak{k}_{\mathbb{C}}$ *by the Cayley transform* [13].

8.3. Root system of type F. A final run through the classification gives

Theorem. *Let* G *be a simple Lie group, and let* G' *be a real form of* G *if* G *is complex, a real form of* $G_{\mathbb{C}}$ *if* G *is absolutely simple. Suppose that* G' *has restricted root system*

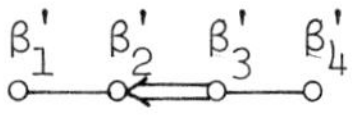

of type F_4.

1. *If* G *has a parabolic subgroup* $P = NAM$ *corresponding to* $P_{\{\beta_1'\}} \subset G'$, *then* N *has square integrable representations and is defined through the action of* $\mathfrak{m}$ *on* $\mathfrak{n}$ *by spin and vector representations.*

2. *If* G *has a parabolic subgroup* $P = NAM$ *corresponding to* $P_{\{\beta_4'\}} \subset G'$, *then* N *is a Heisenberg group with composition specified by the bilinear invariant of* M *on* $\mathfrak{n}/\mathfrak{z}$.

§9. Framework for Fourier Inversion

We indicate how the method of [14] applies to give Fourier inversion formulae for most of the parabolics $P = MAN \subset G$, where G is simple and N has square integrable representations. Then in §§10-17 we run through the classification and give explicit Plancherel formulae. As implicitly just indicated, now MAN denotes a general parabolic under consideration.

The parabolics we consider are those for which the action of M on the center Z of N satisfies

(9.1) there is a nonconstant M-invariant polynomial on $\mathfrak{z}^*$.

We will give a usable criterion for (9.1) at the end of this section. For example [14 , Lemma 5.3] the Pfaffian polynomial $\mathcal{P}(\lambda)$ on $\mathfrak{z}^*$ described in §2 has $|\mathcal{P}|$ invariant under M; so $\mathcal{P}(\lambda)^2$ is an invariant polynomial of degree $\dim_{\mathbb{R}} N/Z$, thus nonconstant if $Z \neq N$.

Let us rephrase (9.1) as the existence of some nonzero

$$\text{(9.2a)} \quad \begin{cases} \psi : \text{real polynomial on } \mathfrak{z}^*, \text{ homogeneous of some} \\ \text{degree } d > 0, \text{ where } |\psi| \text{ is } M\text{-invariant.} \end{cases}$$

That defines

$$\text{(9.2b)} \quad \begin{cases} \Psi : \text{differential operator on } Z, \text{ homogeneous of} \\ \text{degree } d > 0, \text{ where } |\Psi| \text{ is } M\text{-invariant} \end{cases}$$

through Fourier transform $\mathcal{F}: L^2(\mathfrak{z}^*) \to L^2(Z)$, $\mathcal{F}(\varphi) = \hat{\varphi}\cdot\exp$ where $\hat{\varphi} \in L^2(\mathfrak{z})$. Details later. First, set

(9.3a) $k = \dim_{\mathbb{R}} Z, \quad \ell = \dim_{\mathbb{R}} N/Z \quad \text{and} \quad q = k + \frac{1}{2}\ell$,

and consider the Ad(MA)-invariant C^∞ manifold splittings

(9.3b) $N \approx Z \times X \quad \text{and} \quad P = Z \times W, \quad W = X \times M \times A.$

Let the operator Ψ act on the Z variable only. That defines a 1-parameter semigroup $\{D^t\}_{t \geqslant 0}$ of invertible positive self-adjoint operators on $L^2(P)$:

(9.4) $(D^t F)(z,w) = (|\Psi|^{tq/d} F)(z,w) = \mathcal{F}(|\psi|^{tq/d} \cdot \mathcal{F}^{-1}(F))(z,w).$

where $\mathcal{F}$ acts on the $\mathfrak{z}^*$, $\mathfrak{z}$, Z variables only. The D^t, $t > 0$, are unbounded but have a common dense domain consisting of all $F \in C^\infty(P)$ such that

(9.5)
(i) $F(z,\cdot)$ is in the Schwartz space of W, for all $z \in Z$
(ii) $F(\cdot,w)$ is in the Schwartz space of Z, for all $w \in W$
(iii) $(\mathcal{F}^{-1}F)(\cdot,w)$ vanishes to sufficiently high order on the zeroes of ψ, uniformly in w.

See [14]. The Fourier Inversion Formula on P will be of the form

(9,6) $F(1_P) = \int_{\hat{P}} \text{trace } \pi(DF) \, d\mu(\pi)$

where μ is a regular Borel measure on $\hat{P}$. Here D, which is homogeneous of degree q in the Z variable, compensates lack of unimodularity in P.

Let us be more explicit about the integral (9.6) by means of the Mackey machine. The square integrable classes $[\pi_\lambda] \in \hat{N}$ are parameterized by

(9.7a) $$\Lambda = \{\lambda \in \mathfrak{z}^*: \mathcal{P}(\lambda) \neq 0\} .$$

Let $b \mapsto \lambda_b$ denote a Borel section to $\Lambda \to M\backslash\Lambda$. In other words, we write Λ as disjoint union of the orbits

(9.7b) $$\Lambda_b = \mathrm{Ad}^*(M)\cdot\lambda_b .$$

In view of (9.1), the MA-stabilizer of $[\pi_{\lambda_b}]$ is M_bA_1 where

(9.8) $$M_b = \{m \in M: \mathrm{Ad}^*(m)\lambda_b = \lambda_b\} \text{ and } A_1 = \{a \in A: \mathrm{Ad}(a)|_{\mathfrak{z}} = 1\} .$$

It will turn out that these groups are reductive.

Choose a Borel field $\{(\tilde{\pi}_b, \alpha_b)\}$ where $\alpha_b: M_bA_1 \times M_bA_1 \to \mathbb{C}'$ is a Borel cocycle and $\tilde{\pi}_b$ is an α_b^{-1}-representation of NM_bA_1 with $\tilde{\pi}_b|N = \pi_{\lambda_b}$. Then the classes in $\hat{P}$ that correspond to Λ_b and enter into (9.6) are those of the unitarily induced

(9.9) $$\pi_{b,\nu} = \mathrm{Ind}_{NM_bA_1 \uparrow P}(\tilde{\pi}_b \otimes \nu), \quad \nu \in (M_bA_1)^\wedge_{\alpha_b} .$$

More precisely, let S denote the "unit sphere,"

$$S = \{\lambda \in \Lambda: |\psi(\lambda)| = 1\} .$$

Then in each case S will be a finite union of M-orbits,

$$(9.10) \qquad S = S_1 \cup \ldots \cup S_p, \text{ disjoint}, \quad S_i = \Lambda_{b_i} .$$

Set

$$(9.11) \qquad M_i = M_{b_i}, \quad \alpha_i = \alpha_{b_i}, \quad \pi_{i,\nu} = \pi_{b_i,\nu} .$$

Appropriate normalizations of Haar measure on the M_iA_1, thus normalizations of the

$$(9.12a) \qquad \mu_i : \text{ Plancherel measure for } \alpha_i\text{-representations of } M_iA_1,$$

will then reduce (9.6) to an explicit formula

$$(9.12b) \qquad F(1_p) = \sum_{1 \leqslant i \leqslant p} \int_{(M_iA_1)^{\wedge}_{\alpha_i}} \text{trace } \pi_{i,\nu}(DF)\, d\mu_i(\nu).$$

In most cases each $\alpha_i = 1$ so the integration (9.12) will be integration against ordinary Plancherel measure of reductive Lie groups.

Before proceding, let us settle the question (9.1) of existence of nonconstant M-invariant polynomials on $\mathfrak{z}^*$.

<u>9.13. Lemma</u>. <u>Let</u> G <u>be a real split simple Lie group. Then there is a one-to-one correspondence between conjugacy classes of parabolic subgroups</u> $P = MAN \subset G$ <u>with</u> N <u>commutative, and holomorphic equivalence classes of bounded symmetric domains</u> $D = G^{\ddagger}/K^{\ddagger}$ <u>where</u> $G^{\ddagger}$

is a real form of $G_{\mathbb{C}}$. *The correspondence is:* $G_{\mathbb{C}}/P_{\mathbb{C}}$ *is the compact dual of* D.

Proof. Each is defined by a simple root ψ_s whose coefficient $m_s = 1$ in the maximal root $\mu = \Sigma m_i \psi_i$. See [23] or [29 ,§8] or [25].

q.e.d.

9.14. *Lemma.* *Let* $D = G/K$ *be an irreducible bounded symmetric domain, and let* $K' = [K,K]$, *the semisimple part of* K. *Then there is a nonconstant* K'*-invariant complex polynomial on the holomorphic tangent space of* D *if, and only if,* D *is of tube type (holomorhically equivalent to a tube domain over a self-dual cone-see* [13]).

Proof. View D in the Harish-Chandra embedding in its holomorphic tangent space, which we denote by $\mathfrak{v}$ because the usual symbols are playing other roles in this paper. Now K acts on $\mathfrak{v}$ as a subgroup of the (compact) unitary group. As in [13] we write $\check{S}$ for the Bergman-Shilov boundary of D in $\mathfrak{v}$, and express $\check{S} = K/L$ where $c \in G_{\mathbb{C}}$ is the Cayley transform and L is the isotropy subgroup of K at $c(0) \in \check{S}$.

First suppose that D is of tube type. Then Korányi and I [13] gave its realization as a tube domain

$$c(D) = Ad(c)G \cdot c(0)$$

inside $\mathfrak{v}$, as follows. One has a real form $\mathfrak{f} = \mathfrak{v} \cap Ad(c)^2\mathfrak{v}$ of $\mathfrak{v}$ such that $c(0) \in \sqrt{-1}\,\mathfrak{f}$. $Ad(c)^2$ preserves $\mathfrak{k}$, and has order 2, so $\mathfrak{k} = \mathfrak{l} + \mathfrak{q}$ sum of 1 and (-1) eigenspaces of $Ad(c)^2$. Here $\mathfrak{l}$ is the Lie algebra of L. Let K^* be the analytic subgroup of $K_{\mathbb{C}}$ for $\mathfrak{l} + \sqrt{-1}\,\mathfrak{q}$. Then $Ad(K^*)\cdot c(0) = K^*/L$ is a self-dual cone in $\sqrt{-1}\,\mathfrak{J}$ that is the noncompact dual symmetric space of $\check{S}$. Further, $\mathfrak{J}$ has a natural structure of formally real simple Jordan algebra such that $L = Aut(\mathfrak{J})$ and $Ad(K^*)\cdot c(0)$ is the interior of $\sqrt{-1}\,\{u^2 : u \in \mathfrak{J}\}$. The tube domain structure is

$$c(D) = \{x + y \ : x \in \mathfrak{J},\ y \in Ad(K)^*\cdot c(0)\}.$$

Now the point is that $Ad([K^*, K^*])$ preserves the Koecher norm function $\underline{\underline{n}}$ of the complex Jordan algebra structure $\mathfrak{J}_{\mathbb{C}}$ of $\mathfrak{v}$, and $\underline{\underline{n}}$ is a nonconstant complex polynomial of degree equal to the symmetric space rank of D. Evidently $\underline{\underline{n}}$ is $Ad([K^*,K^*]_{\mathbb{C}})$-invariant, hence invariant under $Ad(K')$.

Now suppose that D is not of tube type. Following Schmid [20], consider the K-equivariant orhtogonal decomposition

$$L^2(K/L) = L^2(\check{S}) = \mathcal{H} + \mathfrak{M}$$

where $\mathcal{H}$ is spanned by all boundary values of K-finite holomorphic functions on D. Schmid's result in this, the irreducible non-tube case: let $\bar{\partial}_b$ be the restriction to $\check{S}$ of the $\bar{\partial}$ operator on $\mathfrak{v}$, then $\mathcal{H}$ is spanned by $\{f \in C^\infty(S) : \bar{\partial}_b f = 0\}$.

Now let ψ be a K'-invariant complex polynomial on $\mathfrak{v}$ and $f = \psi|_{\check{S}}$; we will prove that ψ is constant. For this we may assume ψ homogeneous, so $|\psi(z)|^2$ is invariant under the center of K. In consequence, $|f(z)|^2$ is K-invariant and hence constant, so $\bar{\partial}_b(f\bar{f}) = 0$. As f is smooth and $\bar{\partial}_b f = 0$ now also $\bar{\partial}_b \bar{f} = 0$, so we have $f, \bar{f} \in \mathcal{H}$ The center of K is a circle group whose character group is of the form $\{\chi_n : n \in \mathbb{Z}\}$ where χ_n gives the action on polynomials of degree n. So $\chi_n(h) = 0$ for $n < 0$ whenever h is the $\check{S}$-restriction of a holomorphic function. Now χ_n, $n < 0$, kills both f and $\bar{f}$. From the latter, $\chi_m(f) = 0$ for $m > 0$. We conclude that f is constant. It follows that ψ is constant.

q.e.d.

Combining Lemmas 9.13 and 9.14, and noting the example just after (9.1), we have

9.15. Theorem. *Let G be a simple Lie group, let P = MAN be a parabolic subgroup such that N has square integrable representations, and let Z be the center of N.*

1. *If N is noncommutative then $\mathfrak{z}^*$ carries nonconstant M-invariant polynomials.*

2. *Suppose that N is commutative, let G_1 be the real split simple group associated to G, let P_1 be the parabolic subgroup of G_1 associated to P, and let D be the bounded symmetric domain corresponding to P_1 as in* Lemma 9.13. *Then $\mathfrak{z}^*$ carries a nonconstant M-invariant polynomial if, and only if, D is of tube type.*

§10. Fourier Inversion Inside Groups of Type A

We are going to apply the Fourier Inversion procedure, sketched in §9, to parabolic subgroups $P = MAN$ is a group G of type A_{n-1}: $\overset{\psi_1}{\circ}\!\!-\!\!\overset{\psi_2}{\circ}\!\!-\cdots-\!\!\overset{\psi_{n-1}}{\circ}$. The parabolics in question derive from the $P_{\{\psi_s\}}$, $1 \leqslant s \leqslant [\frac{n}{2}]$, of (4.5.1)(i), and the $P_{\{\psi_s,\psi_{n-s}\}}$, $1 \leqslant s \leqslant [\frac{n-1}{2}]$, of (4.5.1)(ii). In order to avoid a bit of a mess in describing the groups M and M_i we will in fact look inside

$U(p,q)$	instead of	$SU(p,q)$
$GL'(m;\mathbb{F})$	instead of	$SL(m;\mathbb{F})$

where

$$GL'(m;\mathbb{F}) = \{\gamma \in GL(m;\mathbb{F}) : \ \gamma \text{ preserves Lebesgue measure on } \mathbb{F}^m\}.$$

Then we can pass back to $SU(p,q)$ and $SL(m;\mathbb{F})$.

10.1.1. Lemma. *There is a nonconstant M-invariant polynomial on* $\mathfrak{z}^*$ *if and only if* P *derives either from* $P_{\{\psi_s\}}$ *with* $2s = n$ *or from* $P_{\{\psi_s,\psi_{n-s}\}}$ *with* $1 \leqslant s \leqslant [\frac{n-1}{2}]$.

Proof. It suffices to consider the split case $G = GL'(n;\mathbb{R})$.

Suppose $P = P_{\{\psi_s\}}$ with $2s < n$. We can apply Theorem 9.15 or use the following elementary argument. Here $\mathfrak{z}^* = \mathfrak{n}^* \cong \mathbb{R}^{(n-s)\times s}$ by $\lambda(z) = \text{trace}\,(\lambda z)$, $z \in \mathfrak{z} = \mathbb{R}^{s\times(n-s)}$. Also, $M = GL'(s;\mathbb{R}) \times GL'(n-s;\mathbb{R})$ acts on $\mathfrak{z}$ by $Ad(\alpha,\beta)\colon z \mapsto \alpha z\beta^{-1}$, so it acts on $\mathfrak{z}^*$ by $Ad^*(\alpha,\beta)\cdot\lambda = \beta\gamma\alpha^{-1}$. The set $\Lambda = \{\lambda \in \mathfrak{z}^* :\ \lambda$ has matrix rank $s\}$ is open in $\mathfrak{z}^*$, is $Ad^*(M)$-stable, and contains $\lambda_0 = \begin{pmatrix} I \\ 0 \end{pmatrix}$. Old fashioned row and column operations show: if $\lambda \in \Lambda$ then $r\lambda_0 \in Ad^*(M)\cdot\lambda$

for some $r > 0$. Since $n-s > s$ we have $(I,\beta) \in M$ where $\beta = \begin{pmatrix} r^{-1}I & 0 \\ 0 & \beta' \end{pmatrix}$, and $Ad^*(I,\beta)\colon r\lambda_0 \mapsto \lambda_0$. So $\Lambda = Ad^*(M)\cdot\lambda_0$ is an open M-orbit on $\mathfrak{z}^*$. Thus $\mathfrak{z}^*$ cannot carry a nonconstant M-invariant polynomial.

Suppose $P = P_{\{\psi_s\}}$ with $2s = n$ or $P = P_{\{\psi_s,\psi_{n-s}\}}$ with $2s < n$. Then $\mathfrak{z}^* \cong \mathbb{R}^{s\times s}$ as above, $M \cong GL'(s;\mathbb{R}) \times GL'(n-2s;\mathbb{R}) \times GL'(s;\mathbb{R})$ acts on it by $Ad^*(\alpha,\beta,\gamma)\colon \lambda \mapsto \gamma\lambda\alpha^{-1}$, and $\psi(\lambda) = \det(\lambda)^2$ is M-invariant.

q.e.d.

Combining the lemma with the result of (4.5.1) and §5.1 we have

10.1.2. Theorem. *The following are the cases in which* $P = MAN$ *is a parabolic subgroup of a reductive group* G *of type* A_{n-1}, N *has square integrable representations, and* P *satisfies* (9.1).

1. $G = U(p,q)$, $p + q = n$, *and* P *is isomorphic to*

$$\{\mathrm{Im}\,\mathbb{C}^{s\times s} + \mathbb{C}^{s\times(p-s,q-s)}\}\cdot\{GL(s;\mathbb{C}) \times U(p-s,q-s)\} \tag{10.1.3}$$

with $1 \leqslant s \leqslant \min(p,q)$.

2. $G = GL'(m;\mathbb{F})$ *where* $\mathbb{F}^m = \mathbb{R}^n$, $\mathbb{C}^n$ *or* $\mathbb{Q}^{n/2}$, *and* P *is isomorphic to*

$$\left\{ \begin{pmatrix} a\alpha & x & z \\ 0 & b\beta & y \\ 0 & 0 & c\gamma \end{pmatrix} \;\middle|\; \begin{array}{l} \alpha,\ \gamma \in GL'(s;\mathbb{F}) \text{ and } \beta \in GL'(m-2s;\mathbb{F}) \\ a,\ b,\ c \in \mathbb{R}^+ \text{ with } abc = 1 \\ x \in \mathbb{F}^{s\times(m-2s)},\ y \in \mathbb{F}^{(m-2s)\times s},\ z \in \mathbb{F}^{s\times s} \end{array} \right\} \tag{10.1.4}$$

with $1 \leqslant s \leqslant [\frac{m}{2}]$.

We now look at these cases along the lines of §9.

10.2. $G = U(p,q)$, $p + q = n$, with P given by (10.1.3). This is the case $F = \mathbb{C}$ of [14, Theorem 4.9].

Here $\mathfrak{z}^* \cong \mathrm{Im}\, \mathbb{C}^{s\times s} = \mathfrak{z}$ under $\lambda(z) = \mathrm{Re\ Trace}\ (\lambda z)$. $M = GL'(s;\mathbb{C}) \times U(p-s,q-s)$ acts on $\mathfrak{z}$ by $\mathrm{Ad}(\gamma,g)\colon z \mapsto \gamma z \gamma^*$, so it acts on $\mathfrak{z}^*$ by $\mathrm{Ad}^*(\gamma,g)\colon \lambda \mapsto (\gamma^*)^{-1}\lambda\gamma^{-1}$. The invariant $|\psi(\lambda)|$ with which we work is given by

(10.2.1) $\psi(\lambda) = \det\,(\sqrt{-1}\,\lambda)$, real polynomial of degree $d = s$.

The data k, ℓ, q of (9.3a) are

(10.2.2) $\dim_{\mathbb{R}} Z = s^2$, $\dim_{\mathbb{R}} N/Z = 2s(n-2s)$, $q = s(n-s)$

Thus the positive self-adjoint operator D on $L^2(P)$ is

(10.2.3) $D = |\Psi|^{n-s}$, differential just when $n-s$ is even.

The "unit sphere" $S = \{\lambda \in \mathfrak{z}^*\colon |\psi(\lambda)| = 1\}$ decomposes under M as $S_0 \cup S_1 \cup \ldots \cup S_s$ where

(10.2.4)
$$\begin{cases} S_i = \mathrm{Ad}^*(M)\cdot\begin{pmatrix} \sqrt{-1}\, I_i & 0 \\ 0 & -\sqrt{-1}\, I_{s-i} \end{pmatrix} \\ \quad = \{\lambda \in S\colon \sqrt{-1}\,\lambda \text{ has } i \text{ eigenvalues} < 0, \\ \qquad\qquad s-i \text{ eigenvalues} > 0\}\ . \end{cases}$$

The isotropy subgroup of M at the indicated element $\lambda_i \in S_i$ is

(10.2.5) $$M_i = U(i,s-i) \times U(p-s,q-s)$$

The group A consists of the positive scalars in the GL factor of $MA = GL(s;\mathbb{C}) \times U(p-s,q-s)$, so $A_1 = \{1\}$ and $M_i A_1$ is given by (10.2.5).

The class $[\pi_i] \in \hat{N}$ defined by λ_i extends to a class $[\tilde{\pi}_i] \in (NM_i)^\wedge$ with $\tilde{\pi}_i|_N = \pi_i$. This is in [14], or can easily be seen from the argument of the corresponding fact in §10.3 below. So we have "generic" representations of $P = NAM$ given by

(10.2.6) $$\pi_{i,\nu} = \mathrm{Ind}_{NM_i \uparrow P}(\tilde{\pi}_i \otimes \nu), \quad [\nu] \in \hat{M}_i, \quad 0 \leqslant i \leqslant s.$$

The Fourier Inversion Formula

(10.2.7) $$\begin{cases} F(1_P) = \sum_{0 \leqslant i \leqslant s} \int_{\hat{M}_i} \text{trace } \pi_{i,\nu}(DF) d\mu_i(\nu) \\ \text{for all } C^\infty \text{ functions } F: P \to \mathbb{C} \text{ that satisfy (9.5)} \end{cases}$$

now follows in a standard way from [14, Theorem 4.9].

<u>10.3</u>. $G = GL'(m;\mathbb{F})$, $\mathbb{F} = \mathbb{R}$, $\mathbb{C}$ <u>or</u> $\mathbb{Q}$, <u>with</u> P <u>given by</u> (10.14). The case $m = 3$, $s = 1$ is [14, Proposition 5.20].

Here $\mathfrak{z}^* \cong \mathbb{F}^{s\times s} = \mathfrak{z}$ under $\lambda(z) = \mathrm{Re}\ \mathrm{Trace}(\lambda z)$. $M = GL'(s;\mathbb{F}) \times GL'(m-2s;\mathbb{F}) \times GL'(s;\mathbb{F})$ acts on $\mathfrak{z}$ by $\mathrm{Ad}(\alpha,\beta,\gamma)\cdot z = \alpha z\gamma^{-1}$, so it acts on $\mathfrak{z}^*$ by $\mathrm{Ad}^*(\alpha,\beta,\gamma): \lambda \mapsto \gamma\lambda\alpha^{-1}$. We work with the invariant $|\psi(\lambda)|$ given by

$$(10.3.1) \qquad \psi(\lambda) = \det_{\mathbb{R}}\lambda\ , \quad \text{real polynomial of degree } d = \varepsilon s, \quad \varepsilon = \dim_{\mathbb{R}}\mathbb{F}$$

where

$$(10.3.2) \qquad \begin{cases} \mathbb{F} = \mathbb{R}: & \det_{\mathbb{R}}\lambda = \det\lambda\ , \quad \text{ordinary determinant} \\ \mathbb{F} \neq \mathbb{R}: & \det_{\mathbb{R}}\lambda \text{ is the module of } \lambda \text{ on Lebesgue measure of } \mathbb{F}^s. \end{cases}$$

The data k, ℓ, q of (9.3a) are

$$(10.3.3) \qquad \dim_{\mathbb{R}} Z = s^2\varepsilon\ , \quad \dim_{\mathbb{R}} N/Z = 2s(m-2s)\varepsilon\ , \quad q = s(m-s)\varepsilon\ .$$

Thus the positive self-adjoint operator D on $L^2(P)$ is

$$(10.3.4) \qquad D = |\Psi|^{m-s}, \quad \text{differential except when } \mathbb{F} = \mathbb{R} \text{ with } m-s \text{ odd.}$$

The "unit sphere" $S = \{\lambda \in \mathfrak{z}^* : |\psi(\lambda)| = 1\}$ is just $GL'(s;\mathbb{F})$. If $\lambda \in S$ now $(\lambda, I, I) \in M$ sends λ to I. So M is transitive on S. Its isotropy subgroup at $I \in S$ is

$$(10.3.5) \qquad M_I = \{(\alpha,\beta,\gamma) \in M:\ \alpha = \gamma\} \cong GL'(s;\mathbb{F}) \times GL'(m-2s;\mathbb{F})\ .$$

The group A is given by the a, b, c of (10.1.4), so $A \cong \mathbb{R}^+ \times \mathbb{R}^+$ acting by $(u,v)\colon (z,y,x) \mapsto (uvz, uy, vx)$ when $2s < m$. The subgroup A_1 acting trivially on $\mathfrak{z}$ is $\{(u,u^{-1})\} \cong \mathbb{R}^+$ when $2s < m$, is $\{1\}$ when $2s = m$. View

$$N = \mathbb{F}^{s\times s} + \mathbb{F}^{(m-2s)\times s} + \mathbb{F}^{s\times(m-2s)}$$

with

$$(z,y,x)(z',y',x') = (z+z'+xy',\ y+y', x+x').$$

Then, in general,

$$M_I A_1 \cong GL'(s;\mathbb{F}) \times GL(m-2s;\mathbb{F})$$

acting on N by

$$\text{(10.3.6)} \qquad Ad(\alpha,\beta)\colon (z,y,x) \mapsto (\alpha z \alpha^{-1}, \beta y \alpha^{-1}, \alpha x \beta^{-1}).$$

10.3.7. Lemma. *The class* $[\pi] \in \hat{N}$ *defined by* $I \in \mathfrak{z}^*$ *extends to a class* $[\tilde{\pi}] \in (NM_I A_1)^\wedge$ *with* $\tilde{\pi}|_N = \pi$.

Proof. Let $Q = \{(z,y,x) \in N\colon x = 0\}$ and define $\chi \in \hat{Q}$ by $\chi(z,y,0) = e^{\sqrt{-1}\,\mathrm{Re\,Trace}(z)}$. Then $\pi = \mathrm{Ind}_{Q\uparrow N}(\chi)$. Note that $\tilde{\chi}\colon QM_I A_1 \to \mathbb{C}$ by $\tilde{\chi}(z,y,0;\alpha,\beta) = e^{\sqrt{-1}\,\mathrm{Re\,Trace}(z)}$ is a well-defined unitary character. Set $\tilde{\pi} = \mathrm{Ind}_{QM_I A_1 \uparrow NM_I A_1}(\tilde{\chi})$. Then $\tilde{\pi}|_N = \pi$.

q.e.d.

Now the "generic" representations of $P = NAM$ are the

$$\pi_\nu = \mathrm{Ind}_{NM_I A_1 \uparrow P} (\tilde{\pi} \otimes \nu) , \quad [\nu] \in (M_I A_1)^\wedge \quad . \tag{10.3.8}$$

The argument of [14 , §§3 and 4] leads directly to the Fourier Inversion Formula

$$\begin{cases} F(1_P) = \displaystyle\int_{\{GL'(s;\mathbb{F}) \times GL(m-2s;\mathbb{F})\}^\wedge} \mathrm{trace}\ \pi_\nu(DF) d\mu(\nu) \\ \\ \text{for all } C^\infty \text{ functions } F: P \to \mathbb{C} \text{ satisfying (9.5)} \end{cases} \tag{10.3.9}$$

where Haar measure, thus Plancherel measure μ on the reductive group $GL'(s;\mathbb{F}) \times GL(m-2s;\mathbb{F})$, is correctly normalized. If D is a differential operator -- which is the case unless $\mathbb{F} = \mathbb{R}$ with $m-s$ odd -- then we need only require that F be a Schwartz class function on P.

<u>10.4.</u> <u>Remarks.</u> If we use $SU(p,q)$ for G in §10.2, then M shrinks to $GL'(s;C) \times SU(p-s,q-s)$, the M-orbit structure of S is unchanged, and the isotropy subgroups M_i shrink to $U(i,s-i) \times SU(p-s,q-s)$. The inversion formula (10.2.7) is not changed.

If we use $SL(m;\mathbb{F})$ for G in §10.3, then M becomes somewhat unwieldy -- all (α,β,γ) in $GL'(s;\mathbb{F}) \times GL'(m-2s;\mathbb{F}) \times GL'(s;\mathbb{F})$ such that

$\mathbb{F} \neq \mathbb{Q}$: $(\det \alpha)(\det \beta)(\det \gamma) = 1$

$\mathbb{F} = \mathbb{Q}$: no additional condition.

The smaller M remains transitive on S except in the case $\mathbb{F} = \mathbb{R}$, $m = 2s$, where there are two orbits, $S_\pm = \{\lambda \in \mathfrak{z}^* = \mathbb{R}^{s\times s}: \det \lambda = \pm 1\}$ $= Ad^*(M)\cdot\lambda_\pm$ where $\lambda_+ = I$ and $\lambda_- = \text{diag}\,\{-1;1,\ldots,1\}$. There, the isotropy $M_\pm = \{(\alpha,\beta,\gamma) \in M: \gamma = \lambda_\pm\cdot\alpha\cdot\lambda_\pm\}$, and the inversion formula (10.3.9) splits into two pieces. One understands this from the case $\mathbb{F} = \mathbb{R}$, $m = 2$, $s = 1$; in $SL(2;\mathbb{R})$ there, P is the $ax + b$ group, $a > 0$, and $\hat{P}$ has 2 generic elements; in $GL'(2;\mathbb{R})$, P is the $ax+b$ group, $a \neq 0$, and $\hat{P}$ has just one generic element.

10.5. Remarks. In most of the cases -- the ones where N is noncommutative -- we have the option of using the Pfaffian polynomial $\mathcal{P}(\lambda)$ for ψ. It has degree $\frac{1}{2}\dim_{\mathbb{R}} N/Z$.

In the cases $G = U(p,q)$ of §10.2, $\mathcal{P}$ has degree $s(n-2s)$, $n = p+q > 2s$. That leads to a positive self adjoint operator of the form $|\Pi|^{(n-s)/(n-2s)}$, differential just when $2(n-2s)$ divides $n-s$, i.e. when there is an integer $b > 0$ such that $(2b-1)n = (4b-1)s$. If that happens then $n \equiv s \pmod 2$ so the operator D of (10.2.3) already is differential.

In the cases $G = GL'(m;\mathbb{F})$ of §10.3, $\mathcal{P}$ has degree $s(m-2s)\varepsilon$, $\varepsilon = \dim_{\mathbb{R}}\mathbb{F}$. That leads to a positive self-adjoint operator of the form $|\Pi|^{(m-s)/(m-2s)}$, differential just when $2(m-2s)$ divides $m-s$, in which case the operator D of (10.3.4) already is differential.

In summary, our choice of operators D maximizes the chance of their being differential, so that the conditions (9.5) can be reduced to "F is Schwartz class" in the inversion formula.

§11. Fourier Inversion Inside Groups of Types B and D

We now apply the Fourier Inversion procedure to parabolic subgroups $P = MAN$ in an orthogonal group G, either

B_n :  with $n \geqslant 2$ or D_n : 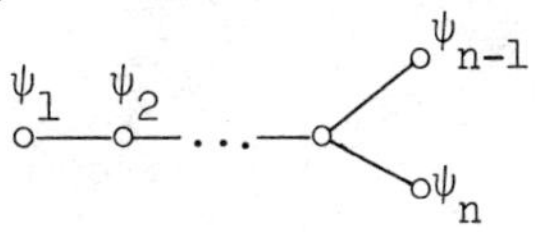

with $n \geqslant 4$. The parabolics derive from

B_n: the $P_{\{\psi_s\}}$ of (4.5.2)(i) with $s = 1$ and $s = 2,4,..,2[\frac{n}{2}]$

D_n: the $P_{\{\psi_s\}}$ of (4.5.4)(i) with $s = 1$ and $s = 2,4,..,2[\frac{n-2}{2}]$

D_n: the $P_{\{\psi_s\}}$ of (4.5.4)(ii) where $s = n-1$ or n

D_n: $P_{\{\psi_{n-1},\psi_n\}}$ of (4.5.4)(iii) where n is odd.

The cases where P has abelian nilradical are $P_{\{\psi_1\}}$ in either B_n or D_n, and $P_{\{\psi_{n-1}\}} \cong P_{\{\psi_n\}}$ in D_n.

<u>11.1.1. Lemma</u>. <u>There is a nonconstant</u> M-<u>invariant polynomial on</u> $\mathfrak{z}^*$ <u>except in the case where</u> $P \subset D_n$ <u>derives from</u> $P_{\{\psi_{n-1}\}}$ <u>or</u> $P_{\{\psi_n\}}$ <u>with</u> n <u>odd</u>.

<u>Proof</u>. If N is noncommutative then the Pfaffian provides the invariant polynomial $\psi(\lambda) = \mathbf{P}(\lambda)^2$, as described just after (9.1).

Now suppose N abelian. We can apply Theorem 9.15 or use the following elementary arguments. We may assume G split, i.e. $G = O(p,q)$ where (p,q) is $(n,n+1)$ or (n,n), $n = [\frac{1}{2}(p+q)]$. If $P = P_{\{\psi_1\}}$ then $N\cdot M = \mathbb{R}^{p-1,q-1}\cdot\{O(p-1,q-1) \times\{\pm 1\}\}$ and the scalar product on $\mathbb{R}^{p-1,q-1}$ defines the invariant polynomial on $\mathfrak{z}^*$.

Finally suppose $P = P_{\{\psi_n\}} \subset O(n,n) = G$. Then $N = \mathrm{Skew}\, \mathbb{R}^{n\times n}$, commutative, and $M = GL'(n;\mathbb{R}) \times \{\pm 1\}$ acting by $\mathrm{Ad}(\gamma, \pm 1)\cdot z = \gamma z \cdot {}^t\gamma$. Identify $\mathfrak{z}^* \cong \mathrm{Skew}\, \mathbb{R}^{s\times s}$ by $\lambda(z) = \mathrm{trace}(\lambda z)$; then M acts on $\mathfrak{z}^*$ by $\mathrm{Ad}^*(\gamma,\pm 1)\colon \lambda \mapsto {}^t\gamma^{-1}\cdot\lambda\cdot\gamma^{-1}$. If n is even then $\psi(\lambda) = \det \lambda$ is the M-invariant polynomial. If n is odd then

$$\Lambda = \{\lambda \in \mathfrak{z}^*\colon \ \lambda \text{ has matrix rank } 2[\tfrac{n}{2}]\}$$

is seen as follows to be an open orbit. Let $\lambda \in \Lambda$. Some $(\gamma,\pm 1) \in M$ carries it to a matrix

$$T(a_1,\ldots,a_m) = \mathrm{diag}\left\{\begin{pmatrix} 0 & -a_1 \\ a_1 & 0 \end{pmatrix},\ldots,\begin{pmatrix} 0 & -a_m \\ a_m & 0 \end{pmatrix}, 0\right\}$$

where $n = 2m+1$ and each $a_j > 0$, for that is a positive Weyl chamber in a Cartan subalgebra of $\mathfrak{so}(n) = \mathrm{Skew}\, \mathbb{R}^{n\times n}$. Now, $(\gamma,1) \in M$ given by $\gamma = \mathrm{diag}\,\{\sqrt{a_1}, \sqrt{a_1};\ldots; \sqrt{a_m}, \sqrt{a_m}; (a_1\ldots a_m)^{-1}\}$ further carries $T(a_1,\ldots,a_m)$ to $T(1,\ldots,1)$. So the open set $\Lambda = \mathrm{Ad}^*(M)\cdot T(1,\ldots,1)$, which precludes existence of nonconstant invariant polynomials.

<u>q.e.d.</u>

Combining the lemma with the results of (4.5.2), (4.5.4) and §§5.2 and 5.4, we have

<u>11.1.2 Theorem</u>. *The following are the cases in which* $P = MAN$ *is a parabolic subgroup of a simple group* G *of type* B_n *or* D_n, N *has square integrable representations, and* P *satisfies* (9.1).

1. $G = O(p,q)$, $1 \leqslant p \leqslant q$ *and* $n = [\frac{1}{2}(p+q)]$, *and* P *is isomorphic to*

(11.1.3) $\{\text{Skew}\, \mathbb{R}^{s\times s} + \mathbb{R}^{s\times(p-s,q-s)}\}\cdot\{GL(s;\mathbb{R}) \times O(p-s,q-s)\}$

where (i) $s = 1$ *or* (ii) $s = 2,4,\ldots, \min(2[p/2], 2[q/2])$.

2. $G = O(m;\mathbb{C})$, $n = [m/2]$, *and* P *is isomorphic to*

(11.1.4) $\{\text{Skew}\, \mathbb{C}^{s\times s} + \mathbb{C}^{s\times(m-2s)}\}\cdot\{GL(s;\mathbb{C}) \times O(m-2s;\mathbb{C})\}$

where (i) $s= 1$ *or* (ii) $s = 2,4,\ldots,2[n/2]$.

3. $G = SO^*(2n)$ *and* P *is isomorphic to*

(11.1.5) $\{\mathfrak{so}^*(2s) + \mathbb{Q}^{s\times(n-2s)}\}\cdot\{GL(s;\mathbb{Q}) \times SO^*(2(n-2s))\}$

where $1 \leqslant s \leqslant [n/2]$.

We run through these cases along the lines sketched in §9.

11.2. $G = O(p,q)$, $1 \leqslant p \leqslant q$, *with* P *given by* (11.1.3)(i). This is [14, (4.19)].

Set $u = p - 1$, $v = q - 1$ and $w = u+v = p+q - 2$. Then $N = \mathbb{R}^{u,v}$, so $\mathfrak{z}^* \cong \mathbb{R}^{u,v}$ under $\lambda(z) = \langle \lambda,z \rangle$, and $M = \{\pm 1\} \times O(u,v)$ acts on $\mathfrak{z}^*$ by $Ad^*(\pm 1,g) = \pm g\lambda$. The polynomial invariant is

$$\psi(\lambda) = \langle \lambda,\lambda \rangle = \sum_1^u \lambda_i^2 - \sum_{u+1}^{u+v} \lambda_i^2 . \tag{11.2.1}$$

As $\dim_{\mathbb{R}} Z + \frac{1}{2}\dim_{\mathbb{R}} N/Z = u+v$, the positive self adjoint operator D on $L^2(P)$ is

$$D = |\Box|^{(u+v)/2}, \quad \Box = \sum_1^u \partial^2/\partial z_i^2 - \sum_{u+1}^{u+v} \partial^2/\partial z_i^2 . \tag{11.2.2}$$

It is differential just when $u+v \equiv 0 \mod 4$ or $u+v = 1$.

The "unit sphere" $S = \{\lambda \in \mathbb{R}^{u,v} : \langle \lambda,\lambda \rangle = \pm 1\}$ decomposes under M as $S_+ \cup S_-$ where $S_\pm$ is given by $\langle \lambda,\lambda \rangle = \pm 1$. Here the isotropy subgroups are

$$M_+ \cong \{\pm 1\} \times O(u-1,v) \quad \text{and} \quad M_- \cong \{\pm 1\} \times O(u,v-1) \tag{11.2.3}$$

at $\lambda_+ : z \mapsto z_1$ and $\lambda_- : z \mapsto z_{u+v}$. Note that $\lambda_\pm$ defines the unitary character $\chi_\pm \in (\mathbb{R}^{u,v})^\wedge$,

$$\chi_+(z) = e^{\sqrt{-1}\, z_1} \quad \text{and} \quad \chi_-(z) = e^{\sqrt{-1}\, z_{u+v}} . \tag{11.2.4}$$

Those extend to $\mathbb{R}^{u,v} \cdot M_\pm$ by $\tilde{\chi}(z,m) = \chi(z)$, so the "generic" representations of P are the

$$\pi_{\pm,\nu} = \mathrm{Ind}_{N \cdot M_\pm \uparrow P}(\tilde{\chi} \otimes \nu), \quad [\nu] \in \widehat{M_\pm} . \tag{11.2.5}$$

As in [14, §§3 and 4] this leads to the Fourier Inversion formula

$$(11.2.6)\quad \begin{cases} F(1_P) = \displaystyle\int_{(\{\pm 1\}\times O(u-1,v))^\wedge} \text{trace } \pi_{+,\nu}(DF)d\mu_+(\nu) \\ \qquad + \displaystyle\int_{(\{\pm 1\}\times O(u,v-1))^\wedge} \text{trace } \pi_{-,\nu}(DF)d\mu_-(\nu) \\ \text{for all } F \in C^\infty(P) \text{ that satisfy (9.5).} \end{cases}$$

11.3. $G = O(p,q)$, $1 \leqslant p \leqslant q$, with P given by (11.1.3)(ii). This is the case $\mathbb{F} = \mathbb{R}$ of [14, Theorem 4.9].

Here $Z = \text{Skew } \mathbb{R}^{s\times s} = \mathfrak{z}$, so $\mathfrak{z}^* \cong \text{Skew } \mathbb{R}^{s\times s}$ under $\lambda(z) = \text{trace }(\lambda z)$. $M = GL'(s;\mathbb{R}) \times O(p-s,q-s)$ acts on $\mathfrak{z}$ by $Ad(\gamma,g)\colon z \mapsto \gamma z\cdot{}^t\gamma$, so it acts on $\mathfrak{z}^*$ by $Ad^*(\gamma,g)\colon \lambda \mapsto {}^t\gamma^{-1}\cdot\lambda\cdot\gamma^{-1}$. The polynomial invariant we use is the Pfaffian of λ relative to a volume element in $\mathbb{R}^s$,

$$(11.3.1)\qquad \psi(\lambda) = \text{Pfaffian }(\lambda), \quad \text{degree } d = s/2.$$

In terms of the integer $d = s/2$, the data k, ℓ, $k + \frac{1}{2}\ell$ of (9.3a) are $2d^2 - d$, $2d(p+q-4d)$, $d(p+q-2d-1)$. Thus the positive self-adjoint operator D on $L^2(P)$ is

$$(11.3.2)\qquad D = |\psi|^{p+q-s-1}, \quad \text{differential just when } p+q-s \text{ is odd.}$$

The "unit sphere" $S = \{\lambda \in \mathfrak{z}^*\colon |\det \lambda| = 1\}$ because $|\text{Pfaffian}|^2 = |\det|$. As in the last part of the proof of Lemma 11.1.1, M is transitive on S. The isotropy subgroup at $J = \begin{pmatrix} 0 & I \\ -I & 0 \end{pmatrix}$, $d \times d$ blocks with $s = 2d$, is

(11.3.3) $$M_J = Sp(s/2;\mathbb{R}) \times O(p-s,q-s).$$

The group A consists of the positive scalars in the GL factor of $MA = GL(s;\mathbb{R}) \times O(p-s,q-s)$, so its subgroup $A_1 = \{a \in A: Ad(a)|_{\mathfrak{z}} \text{ is trivial}\} = \{1\}$. Now $M_J A_1$ is given by (11.3.3).

The class $[\pi] \in \hat{N}$ defined by $J \in \mathfrak{z}^*$ extends to a class $[\tilde{\pi}] \in (NM_J)^{\wedge}$ by a refinement [28, pp. 52ff] of the method of Lemma 10.3.7. So the "generic" representations of P are the

(11.3.4) $$\pi_\nu = \mathrm{Ind}_{NM_J \uparrow P}(\tilde{\pi} \otimes \nu), \quad [\nu] \in \widehat{M_J}.$$

Now the Fourier Inversion formula

(11.3.5) $$\begin{cases} F(1_P) = \displaystyle\int_{\{Sp(s/2;\mathbb{R})\times O(p-s,q-s)\}^{\wedge}} \operatorname{trace} \pi_\nu(DF)d\mu(\nu) \\ \text{for all } F \in C^\infty(P) \text{ that satisfy (9.5)} \end{cases}$$

follows by standard technique from [14, Theorem 4.9].

<u>11.4.</u> $G = O(m;\mathbb{C})$ <u>with</u> P <u>given by</u> (11.1.4)(i). This is the complexification of the case studied in §11.2.

Here $N = Z = \mathbb{C}^{m-2}$ and $\mathfrak{z}^* \cong \mathbb{C}^{m-2}$ under $\lambda(g) = \mathrm{Re} \sum_1^{m-2} \lambda_i z_i$. $M = U(1) \times O(m-2;\mathbb{C})$ acts on $\mathfrak{z}^*$ by $Ad^*(\gamma,g): \lambda \mapsto \bar{\gamma}g\lambda$, which has $\mathbb{R}$-polynomial invariant

(11.4.1) $$\psi(\lambda) = |\langle\lambda,\lambda\rangle|^2 = |\sum_1^{m-2} \lambda_i^2|^2, \quad \text{degree } 4.$$

As $\dim_{\mathbb{R}} Z + \frac{1}{2}\dim_{\mathbb{R}} N/Z = 2(m-2)$, the positive self adjoint operator D on $L^2(P)$ is

$$(11.4.2) \qquad D = \Psi^{(m-2)/2}, \quad \Psi = \sum_{i,j=1}^{m-2} \partial^4/\partial z_i^2 \partial \bar{z}_j^2.$$

It is differential just when m is even.

The "unit sphere" $S = \{\lambda \in \mathfrak{z}^* : |\langle \lambda,\lambda\rangle| = 1\}$ is homogeneous under M: given $\lambda_1 \in S$, if $\lambda \in S$ some $g \in O(m-2;\mathbb{C})$ carries λ to $e^{\sqrt{-1}\theta}\lambda_1$ where $\langle \lambda,\lambda\rangle = e^{2\sqrt{-1}\theta}\langle \lambda_1,\lambda_1\rangle$, and then $e^{\sqrt{-1}\theta} \in U(1)$ carries it on to λ_1. The isotropy subgroup is easily seen to be

$$(11.4.3) \qquad M_1 \cong \{\pm 1\} \times O(m-3;\mathbb{C}).$$

The unitary character $\chi(z) = e^{\sqrt{-1}\,\mathrm{Re}\,\langle \lambda_1, z\rangle}$ on Z for λ_1 extends to ZM_1 by $\tilde{\chi}(z,m) = \chi(z)$, and P has "generic" representations

$$(11.4.4) \qquad \pi_\nu = \mathrm{Ind}_{NM_1 \uparrow P}(\tilde{\chi} \otimes \nu), \quad [\nu] \in \widehat{M_1}.$$

As in [14, §§3 and 4], we end up with the Fourier Inversion Formula

$$(11.4.5) \qquad \begin{cases} F(1_P) = \displaystyle\int_{(\{\pm 1\}\times O(m-3;\mathbb{C}))^\wedge} \mathrm{trace}\, \pi_\nu(DF)\,d\mu(\nu) \\ \text{for } F \in C^\infty(P) \text{ that satisfy (9.5).} \end{cases}$$

<u>11.5</u>. $G = O(m;\mathbb{C})$ <u>with</u> P <u>given by</u> (11.1.4)(ii). This is the complexification of the case studied in §11.3.

Here $Z = \text{Skew}\ \mathbb{C}^{s\times s} = \mathfrak{z}$, so $\mathfrak{z}^* \cong \text{Skew}\ \mathbb{C}^{s\times s}$ under $\lambda(z) = \text{Re Trace}\ (\lambda z)$. $M = GL'(s;\mathbb{C}) \times O(m-2s;\mathbb{C})$ acts on $\mathfrak{z}$ by $Ad(\gamma,g)\colon z \mapsto \gamma z \cdot {}^t\gamma$, hence acts on $\mathfrak{z}^*$ by $Ad^*(\gamma,g)\colon\lambda \mapsto {}^t\gamma^{-1}\cdot\lambda\cdot\gamma^{-1}$. We use the $\mathbb{R}$-polynomial invariant that derives from the (complex) Pfaffian of λ on $\mathbb{C}^s$,

$$\psi(\lambda) = |\text{Pfaffian}(\lambda)|^2, \quad \text{degree } d = s. \tag{11.5.1}$$

The data k,ℓ, q of (9.3a) are $s(s-1)$, $2s(m-2s)$ and $s(m-s-1)$, so the positive self-adjoint operator D on $L^2(P)$ is

$$D = \Psi^{m-s-1}, \quad \text{differential operator.} \tag{11.5.2}$$

The "unit sphere" $S = \{\lambda \in \mathfrak{z}^*\colon |\det \lambda| = 1\}$, and M carries any $\lambda \in S$ to the normal form $J = \begin{pmatrix} 0 & I \\ -I & 0 \end{pmatrix} \in S$, $s/2$ by $s/2$ blocks. The isotropy subgroup

$$M_J = Sp(s/2;\mathbb{C}) \times O(m-2s;\mathbb{C}), \tag{11.5.3}$$

and the class $[\pi] \in \hat{N}$ defined by $J \in \mathfrak{z}^*$ extends to a class $[\tilde{\pi}] \in (NM_J)^\wedge$ by [28 , pp. 52ff]. So P has "generic" representations

$$\pi_\nu = \text{Ind}_{NM_J\uparrow P}\ (\tilde{\pi} \otimes \nu), \quad [\nu] \in \widehat{M_J}. \tag{11.5.4}$$

As in [14 ,§§3 and 4], this results in the Fourier Inversion Formula

$$(11.5.5)\quad \begin{cases} F(1_P) = \int_{\{Sp(s/2;\mathbb{C})\times O(m-2s;\mathbb{C})\}^\wedge} \text{trace } \pi_\nu(DF)d\mu(\nu) \\ \text{for every Schwartz class function } F \text{ on } P. \end{cases}$$

11.6. $G = SO^*(2n)$ and P is given by (11.1.5). The case just studied in §11.5, with $m = 2n$, is the complexification of this case.

We have $\mathfrak{z}^* \cong \mathfrak{so}^*(2s) = \mathfrak{z}$ by $\lambda(z) = \text{Re Trace }(\lambda z)$, where $\mathfrak{so}^*(2s) \subset \mathbb{Q}^{s\times s}$ is the infinitesimal stabilizer of the nondegenerate skew-hermitian form $[u,v] = \sum u_a i\bar{v}_a$ on $\mathbb{Q}^s$. See [28] and [30 ,§8]. $M = GL'(s;\mathbb{Q}) \times SO^*(2(n-2s))$ acts on $\mathfrak{z}$ and $\mathfrak{z}^*$ by

$$Ad(\gamma,g)\colon z \mapsto -\gamma z i\cdot {}^t\bar{\gamma} i \quad\text{and}\quad Ad^*(\gamma,g)\colon \lambda \mapsto -i\,{}^t\bar{\gamma}^{-1} i\lambda\gamma^{-1} \quad .$$

Our $\mathbb{R}$-polynomial invariant derives from the complex Pfaffian of $\lambda \in \mathfrak{so}^*(2s) \subset \mathfrak{o}(2s;\mathbb{C}) \subset \mathbb{C}^{2s\times 2s}$,

$$(11.6.1)\qquad \psi(\lambda) = |\text{Pfaffian of } \lambda \text{ on } \mathbb{C}^{2s}|^2 , \quad \text{degree } d = 2s.$$

The data k, ℓ, q of (9.3a) are $2s^2-s$, $4s(n-2s)$ and $2s(n-s-\frac{1}{2})$, so the positive self-adjoint operator on $L^2(P)$ is

$$(11.6.2)\qquad D = \Psi^{n-s-1/2}, \quad \text{never differential.}$$

The "unit sphere" $S = \{\lambda \in \mathfrak{z}^*\colon \det_{\mathbb{R}}\lambda = 1\}$, or in other words $S = \{\lambda \in \mathbb{Q}^{s\times s}\colon {}^t(\overline{i\lambda}) = i\lambda \text{ and } \det_{\mathbb{R}}(i\lambda) = 1\}$. The action of M in terms of $i\lambda$ is $Ad^*(\gamma,g)\colon i\lambda \to {}^t\bar{\gamma}^{-1}\cdot i\lambda\cdot\gamma^{-1}$. This is just the action of $GL'(s;\mathbb{Q})$ on unimodular hermitian forms on $\mathbb{Q}^s$: there are $s+1$ orbits, those of the $i\lambda_a = \begin{pmatrix} -I_a & 0 \\ 0 & I_{s-a} \end{pmatrix}$ for $0 \leqslant a \leqslant s$. Thus S decomposes under M as $S_0 \cup S_1 \cup \ldots \cup S_s$ where

$$S_a = Ad^*(M)\cdot\lambda_a\ ,\ \lambda_a = \begin{pmatrix} iI_a & 0 \\ 0 & -iI_{s-a} \end{pmatrix},\quad 0 \leqslant a \leqslant s. \tag{11.6.3}$$

The isotropy subgroup of M at λ_a is

$$M_a = Sp(a,s-a) \times SO^*(2(n-2s)). \tag{11.6.4}$$

The group $A_1 = \{1\}$, so $M_aA_1 = M_a$ given by (11.6.4).

The class $[\pi_a] \in \hat{N}$ defined by $\lambda_a \in \mathfrak{z}^*$ extends to a class $[\tilde{\pi}_a] \in (NM_a)^\wedge$; see [28]. Thus the "generic" representations of P are the

$$\pi_{a,\nu} = Ind_{NM_a \uparrow P}(\tilde{\pi}_a \otimes \nu),\quad [\nu] \in \widehat{M_a}\ . \tag{11.6.5}$$

As in [14 , §§3 and 4], this leads to the Fourier Inversion Formula

$$\text{(11.6.6)}\quad \begin{cases} F(1_P) = \sum_{a=0}^{s} \int_{\{Sp(a,s-a)\times SO^*(2(n-2s))\}^\wedge} \text{trace } \pi_{a,\nu}(DF)d\mu_a(\nu) \\ \text{for all } F \in C^\infty(P) \text{ that satisfy (9.5)}. \end{cases}$$

11.7. Remarks. Let us consider use of the Pfaffian $\mathcal{P}(\lambda)$ described just after (9.1), in place of the polynomials ψ of (11.3.1), (11.5.1) and (11.6.1).

In the cases $G = O(p,q)$ of §11.3, $\mathcal{P}$ has degree $\frac{1}{2}s(p+q-2s)$. That leads to a positive self adjoint operator of the form $|\Pi|^{(p+q-s-1)/(p+q-2s)}$, differential just when $2(p+q-2s)$ divides $p+q-s-1$. But then $p+q-s = 2b(p+q-2s)+1$ for some integer $b > 1$, so $p+q-s$ is odd and the operator D of (11.3.2) already is differential.

In the cases $G = O(m;\mathbb{C})$ of §11.5, $\mathcal{P}$ has degree $s(m-2s)$. That leads to a positive self adjoint operator of the form $|\Pi|^{(m-s-1)/(m-2s)}$, not often differential, while the operator D of (11.5.2) always is differential.

In the cases $G = SO^*(2n)$ of §11.6, $\mathcal{P}$ has degree $2s(n-2s)$. That leads to a positive self adjoint operator of the form $|\Pi|^{(n-s-1/2)/(n-2s)}$, never differential, just as the operator D of (11.6.2).

§12. Fourier Inversion Inside Groups of Type C

We finish the classical cases by applying our Fourier Inversion method to the prabolic subgroups $P = MAN$ in a symplectic group

G: 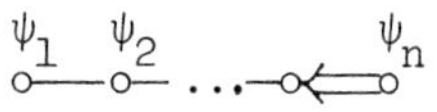 with $n \geqslant 2$. The parabolics in question derive from the $P_{\{\psi_s\}}$, $1 \leqslant s \leqslant n$.

<u>12.1.1. Lemma</u>. *There is a nonconstant M-invariant polynomial on* $\mathfrak{z}^*$.

Proof. We may suppose $G = Sp(n;\mathbb{R})$. Then $\mathfrak{z}^* \cong \mathrm{Sym}\,\mathbb{R}^{s\times s}$ with $M = GL'(s;\mathbb{R}) \times Sp(n-s;\mathbb{R})$ acting by $\mathrm{Ad}^*(\gamma,g)\colon \lambda \mapsto {}^t\gamma^{-1}\cdot\lambda\cdot\gamma^{-1}$. An invariant is $\det(\lambda)$.

q.e.d.

Combining the lemma with the results of (4.5.3) and §5.3 we have

<u>12.1.2. Theorem</u>. *The following are the cases in which $P = MAN$ is a parabolic subgroup in a simple group G of type C_n, N has square integrable representations, and (this is automatic) P satisfies (9.1).*

1. $G = Sp(n;\mathbb{R})$ *and* P *is isomorphic to*

$$\{\mathrm{Sym}\,\mathbb{R}^{s\times s} + \mathbb{R}^{s\times 2(n-s)}\}\cdot\{GL(s;\mathbb{R}) \times Sp(n-s;\mathbb{R})\} \tag{12.1.3}$$

where $1 \leqslant s \leqslant n$.

2. $G = Sp(n;\mathbb{C})$ *and* P *is isomorphic to*

(12.1.4) $$\{\mathrm{Sym}\, \mathbb{C}^{s\times s} + \mathbb{C}^{s\times 2(n-s)}\}\cdot\{GL(s;\mathbb{C}) \times Sp(n-s;\mathbb{C})\}$$

where $1 \leqslant s \leqslant n$.

3. $G = Sp(p,q)$, $1 \leqslant p \leqslant q$ *and* $p + q = n$, *and* P *is isomorphic to*

(12.1.5) $$\{\mathrm{Im}\, \mathbb{Q}^{s\times s} + \mathbb{Q}^{s\times(p-s,q-s)}\}\cdot\{GL(s;\mathbb{Q}) \times Sp(p-s,q-s)\}$$

where $1 \leqslant s \leqslant p$.

Now we run through these three cases.

12.2. $G = Sp(n;\mathbb{R})$ *with* P *given by* (12.1.3).

Here $Z = \mathrm{Sym}\, \mathbb{R}^{s\times s} = \mathfrak{z}$, so $\mathfrak{z}^* \cong \mathrm{Sym}\, \mathbb{R}^{s\times s}$ under $\lambda(z) = \mathrm{trace}\,(\lambda z)$. $M = GL'(s;\mathbb{R}) \times Sp(n-s;\mathbb{R})$ acts on $\mathfrak{z}$ by $Ad(\gamma,g)\colon z \mapsto \gamma z\cdot{}^t\gamma$, hence on $\mathfrak{z}^*$ by $Ad^*(\gamma,g)\colon \lambda \mapsto {}^t\gamma^{-1}\cdot\lambda\cdot\gamma^{-1}$. Our polynomial invariant is

(12.2.1) $$\psi(\lambda) = \det \lambda \quad , \quad \text{degree } d = s.$$

The data k, ℓ, q of (9.3a) are $\frac{1}{2}s(s+1)$, $2s(n-s)$ and $\frac{1}{2}s(2n-s+1)$, so the positive self adjoint operator D on $L^2(P)$ is

(12.2.2) $D = |\Psi|^{(2n-s+1)/2}$, differential just when 4 divides $2n-s+1$.

The "unit sphere" $S = \{\lambda \in \mathfrak{z}^*: |\det \lambda| = 1\}$ decomposes under M as $S_0 \cup S_1 \cup \ldots \cup S_s$ where

$$(12.2.3) \qquad S_i = Ad^*(M)\cdot\lambda_i, \quad \lambda_i = \begin{pmatrix} I_i & 0 \\ 0 & -I_{s-i} \end{pmatrix}, \quad 0 \leqslant i \leqslant s.$$

The isotropy subgroup of M at λ_i is

$$(12.2.4) \qquad M_i = O(i,s-i) \times Sp(n-s;\mathbb{R}) .$$

The group $A_1 = \{1\}$, so M_iA_1 is given by (12.2.4).

Suppose first that $s = n$. Then N is abelian and λ_i defines a unitary character

$$(12.2.5) \qquad \chi_i \in \hat{N} \quad \text{given by} \quad \chi_i(z) = e^{\sqrt{-1}\,\lambda_i(z)} .$$

This character extends to

$$(12.2.6) \qquad \tilde{\chi}_i \in (NM_i)^\wedge \quad \text{given by} \quad \tilde{\chi}_i(z,m) = \chi_i(z).$$

The "generic" representations of P are the

$$(12.2.7) \qquad \pi_{i,\nu} = \mathrm{Ind}_{NM_i \uparrow P}(\tilde{\chi}_i \otimes \nu), \quad [\nu] \in \widehat{M_i}.$$

As before, this leads to the Fourier Inversion formula

$$(12.2.8)\quad \begin{cases} F(1_P) = \sum_{i=0}^{n} \int_{O(i,n-i)^\wedge} \text{trace } \pi_{i,\nu}(DF)d\mu_i(\nu) \\ \text{for all } F \in C^\infty(P) \text{ that satisfy (9.5).} \end{cases}$$

Now suppose that $1 \leqslant s < n$. The class $[\pi_i] \in \hat{N}$ defined by $\lambda_i \in \mathfrak{z}^*$ does not extend to a unitary representation class of NM_i. See [28 , §9], especially Theorem 9.19 there. Instead, π_i extends to a cocycle representation as follows. Write $\mathbb{C}'$ for the circle group $U(1) = GL'(1;\mathbb{C})$. Then the cohomology $H^2(Sp(n-s;\mathbb{R});\mathbb{C}')$ based on Borel cocycles $Sp(n-s;\mathbb{R}) \times Sp(n-s;\mathbb{R}) \to \mathbb{C}'$ is itself a circle group. Denote

(12.2.9) $\bar{\sigma}$: element of order 2 in the circle $H^2(Sp(n-s;\mathbb{R});\mathbb{C}')$,

(12.2.10) $\sigma \in H^2(M_i;\mathbb{C}')$: pullback of $\bar{\sigma}$ under $M_i \to Sp(n-s;\mathbb{R})$.

Then [28, Theorem 9.19] tells us that

(12.2.11) π_i extends to a σ-representation $\tilde{\pi}_i$ of NM_i.

Thus the "generic" representations of P are the

(12.2.12) $\pi_{i,\nu} = \text{Ind}_{NM_i \uparrow P}(\tilde{\pi}_i \otimes \nu)$, $[\nu] \in (M_i)^\wedge_\sigma$

where the last symbol denotes the space of unitary equivalence classes of σ-representations of M_i. Again, the method of [14 , §§3 and 4] gives a Fourier Inversion formula

$$(12.2.13)\qquad \begin{cases} F(1_P) = \sum_{i=0}^{s} \int_{\{O(i,s-i)\times Sp(n-s;\mathbb{R})\}^{\wedge}_{\sigma}} \text{trace}\ \pi_{i,\nu}(DF)d\mu_i(\nu) \\ \text{for all } F \in C^\infty(P) \text{ that satisfy } (9.5) \end{cases}$$

where μ_i is appropriately normalized Plancherel measure on $\{O(i,s-i) \times Sp(n-s;\mathbb{R})\}^{\wedge}_{\sigma}$. If we denote

$$(12.2.14)\qquad \varepsilon : Mp(n-s;\mathbb{R}) \to Sp(n-s;\mathbb{R}) \ , \quad \text{the 2-sheeted cover},$$

then we may view $\{O(i,s-i) \times Sp(n-s;\mathbb{R})\}^{\wedge}_{\sigma}$ as the set of all $[\zeta] \in \{O(i,s-i) \times Mp(n-s;\mathbb{R})\}^{\wedge}$ that are nontrivial on the kernel of ε, and μ_i as the restriction of ordinary Plancherel measure on $\{O(i,s-i) \times Mp(n-s;\mathbb{R})\}^{\wedge}$.

12.3. $G = Sp(n;\mathbb{C})$ with P given by (12.1.4). This is the complexification of the case studied in §12.2.

$Z = \text{Sym}\ \mathbb{C}^{s\times s} = \mathfrak{z}$, $\mathfrak{z}^* \cong \text{Sym}\ \mathbb{C}^{s\times s}$ under $\lambda(z) = \text{Re Trace}\ (\lambda z)$, $M = GL'(s;\mathbb{C}) \times Sp(n-s;\mathbb{C})$ acts on $\mathfrak{z}$ by $Ad(\gamma,g): z \mapsto \gamma z\cdot {}^t\gamma$, and M acts on $\mathfrak{z}^*$ by $Ad^*(\gamma,g): \lambda \mapsto {}^t\gamma^{-1}\cdot\lambda\cdot\gamma^{-1}$. Our polynomial invariant is

$$(12.3.1)\qquad \psi(\lambda) = |\det \lambda|^2, \quad \text{degree} \quad d = 2s.$$

The data k, ℓ, q of (9.3a) are $s(s+1)$, $4s(n-s)$ and $s(2n-s+1)$, so the positive self-adjoint operator D on $L^2(P)$ is

$$D = \Psi^{(2n-s+1)/2}, \text{ differential just when } s \text{ is odd.} \tag{12.3.2}$$

The "unit sphere" $S = \{\lambda \in \mathfrak{z}^*: |\det \lambda| = 1\}$ is homogeneous under M, and the isotropy subgroup of M at $I \in S$ is

$$M_I = O(s;\mathbb{C}) \times Sp(n-s;\mathbb{C}). \tag{12.3.3}$$

The group $A_1 = \{1\}$ so (12.3.3) give $M_I A_1$. The class $[\pi] \in \hat{N}$ defined by $I \in \mathfrak{z}^*$ extends to a class $[\tilde{\pi}] \in (NM_I)^\wedge$; see [28 , Theorem 9.19]. So the "generic" representations of P are the

$$\pi_\nu = \mathrm{Ind}_{NM_I \uparrow P} (\tilde{\pi} \otimes \nu), \quad [\nu] \in \widehat{M_I} . \tag{12.3.4}$$

As in [14, §§3 and 4], this leads to a Fourier Inversion formula

$$\left\{ \begin{array}{l} F(1_P) = \displaystyle\int_{\{O(s;\mathbb{C})\times Sp(n-s;\mathbb{C})\}^\wedge} \mathrm{trace}\ \pi_\nu(DF)d\mu(\nu) \\ \\ \text{for all } C^\infty \ F: P \to \mathbb{C} \text{ that satisfy (9.5) .} \end{array} \right. \tag{12.3.5}$$

If s is odd, so D is differential, F need only be a Schwartz class function on P.

12.4. $G = Sp(p,q)$, $1 \leqslant p \leqslant q$ and $p + q = n$, with P given by (12.1.5). This is the case $\mathbb{F} = \mathbb{Q}$ of [14, Theorem 4.9].

Here $Z = \operatorname{Im} \mathbb{Q}^{s\times s} = \mathfrak{z}$, so $\mathfrak{z}^* \cong \operatorname{Im} \mathbb{Q}^{s\times s}$ under $\lambda(z) = \operatorname{Re} \operatorname{Trace}(\lambda z)$. $M = GL'(s;\mathbb{Q}) \times Sp(p-s,q-s)$ acts on $\mathfrak{z}$ by $\operatorname{Ad}(\gamma,g)\colon z \mapsto \gamma z \gamma^*$, so it acts on $\mathfrak{z}^*$ by $\operatorname{Ad}^*(\gamma,g)\colon \lambda \mapsto (\gamma^*)^{-1}\cdot\lambda\cdot\gamma^{-1}$. Now define a map $h\colon \mathfrak{z}^* \to \operatorname{Im} \mathbb{C}^{2s\times 2s}$ by

$$\mathfrak{z}^* = \operatorname{Im} Q^{s\times s} = \mathfrak{sp}(s) \subset \mathfrak{u}(2s) = \operatorname{Im} C^{2s\times 2s}$$

and set

(12.4.1) $\psi(\lambda) = \det(\sqrt{-1}\, h(\lambda))$, real polynomial of degree $d = 2s$.

The data k, ℓ, $k + \frac{1}{2}\ell$ of (9.3a) are $2s^2 + s$, $4s(p+q-2s)$ and $2sn - 2s^2 + s$, so the positive self adjoint operator D on $L^2(P)$ is

(12.4.2) $D = |\Psi|^{n-s+1/2}$, never differential.

The "unit sphere" $S = \{\lambda \in \mathfrak{z}^*\colon \det_{\mathbb{R}}\lambda = 1\}$ is a single orbit $\operatorname{Ad}^*(M)\cdot J$ where $J = iI$, and the isotropy subgroup there is

(12.4.3) $M_J = SO^*(2s) \times Sp(p-s,q-s)$.

A_1 is trivial so $M_J A_1$ is given by (12.4.3).

The class $[\pi] \in \hat{N}$ defined by $J \in \mathfrak{z}^*$ extends to a class $[\tilde{\pi}] \in (NM_J)^\wedge$; see [28]. The "generic" representations of P now are the

$$\pi_\nu = \mathrm{Ind}_{NM_J \uparrow P} (\tilde{\pi} \otimes \nu), \quad [\nu] \in \widehat{M_J} . \tag{12.4.4}$$

Now [14, Theorem 4.9] one has the Fourier Inversion formula

$$\begin{cases} F(1_P) = \displaystyle\int_{\{SO^*(2s)\times Sp(p-s,q-s)\}^\wedge} \mathrm{trace}\ \pi_\nu(DF)d\mu(\nu) \\ \text{whenever } F \in C^\infty(P) \text{ and satisfies (9.5) .} \end{cases} \tag{12.4.5}$$

§13. Fourier Inversion Inside the Group G_2

We now start on the exceptional groups, applying our Fourier Inversion procedure to parabolic subgroups $P = MAN$ in a group G of type G_2: $\overset{\psi_1}{\circ}\!\Rrightarrow\!\overset{\psi_2}{\circ}$.

The parabolics in question derive from $P_{\{\psi_1\}}$, and satisfy (9.1) because the nilradical is not abelian. From (6.6.1) and §7.1 there are only two cases

(13.1.1) $\quad G = G_{2,A_1A_1}$ and $P = \{\mathbb{R} + \mathbb{R}^4\}\cdot GL(2;\mathbb{R})$,

(13.1.2) $\quad G = (G_2)_{\mathbb{C}}$ and $P = \{\mathbb{C} + \mathbb{C}^4\}\cdot GL(2;\mathbb{C})$.

13.2. Mackey obstructions. In order to work with the exceptional groups we will need a method for computing certain Mackey obstructions, that will replace reference to [27] and [28]. If we start with

(13.2.1a) $\quad M_1$: semisimple linear Lie group and

(13.2.1b) $\quad f\colon M_1 \to Sp(n;\mathbb{R})$ homomorphism

then we want a method to

(13.2.1c) $\quad$ describe $f_\# \pi_1(M_1)$ inside $\pi_1(Sp(n;\mathbb{R}))$.

Once that is done we apply

13.2.2. Proposition. Let c be a generator of the infinite cyclic group $\pi_1(Sp(n;\mathbb{R}))$, and let $r \geqslant 0$ be the integer such that $f_\# \pi_1(M_1)$ is the subgroup generated by c^r. Let $\bar{\sigma}$ be the element of order 2 in the circle group $H^2(Sp(n;\mathbb{R});\mathbb{C}')$ and $\sigma = f^*\bar{\sigma} \in H^2(M_1;\mathbb{C}')$. Then $\sigma = 1$ if and only if r is even.

Proof. We know [28, Lemma 9.17] that $\pi_1(\cdot)^\wedge \cong H^2(\cdot;\ \mathbb{C}')$ in a way that identifies the transpose of $f_\#$ with f^*. Let $a \in \pi_1(M_1)$ with $f_\# a = c^r$ and let $\bar{\tau} \in \pi_1(Sp(n;\mathbb{R}))^\wedge$ correspond to $\bar{\sigma}$. Then $\tau = ({}^t f_\#)(\bar{\tau}) = \bar{\tau} \circ f \in \pi_1(M_1)^\wedge$ corresponds to σ. Now $\sigma = 1$ iff $\tau(a) = 1$ iff $\bar{\tau}(c^r) = 1$ iff $(-1)^r = 1$.

q.e.d.

The question of (13.2.1) immediately reduces to the corresponding question for maximal compact subgroups. So we have

(13.2.3a) $\quad K_1$: compact Lie group

(13.2.3b) $\quad f: K_1 \to U(n)$ homomorphism

and we want to

(13.2.3c) $\quad$ describe $f_\# \pi_1(K_1)$ inside $\pi_1(U(n))$.

If $\pi_1(K_1)$ is finite, then $f_\# \pi_1(K_1)$ is a finite subgroup of the infinite cyclic group $\pi_1(U(n))$, hence trivial. So $r = 0$, even, in Proposition 13.2.2.

Now suppose $\pi_1(K_1)$ infinite. If K_1 is connected, and anyway in all the cases we will consider,

$$(13.2.4) \qquad \begin{cases} K_1 = \{\mathbb{C}' \times K_1'\}/\mathbb{Z}_h & \text{where } \mathbb{Z}_h \text{ is generated} \\ \text{by some } (e^{2\pi i/h},k) & \text{and where } f_{\#}\pi_1(K_1') = \{1\} . \end{cases}$$

In fact K_1' will usually be the semisimple part of K_1. Now consider a commutative diagram

$$(13.2.5) \qquad \left\{ \begin{array}{ccccccc} (e^{i\theta},k) & \in & \mathbb{C}' \times K_1' & \xrightarrow{\tilde{f}} & \mathbb{C}' \times SU(n) & \ni & (e^{i\varphi}, x) \\ \downarrow & & \downarrow \pi & & p \downarrow & & \downarrow \\ e^{im\theta}\cdot k & \in & K_1 & \xrightarrow{f} & U(n) & \ni & e^{i\varphi}\cdot x \end{array} \right.$$

where $(e^{i\theta},k) \mapsto e^{i\theta}\cdot k$ is the usual projection $\mathbb{C}' \times K_1' \to \{\mathbb{C}' \times K_1'\}/\mathbb{Z}_h$ and the modification by the integer m in π is such that the lift $\tilde{f}$ of $f\cdot\pi$ exists.

13.2.6. Proposition. *Suppose that* $\tilde{f}(e^{i\theta},1) = (e^{iq\theta}, A(\theta))$ *for some integer* q, $0 \leqslant \theta \leqslant 2\pi$. *Then* $f_{\#}\pi_1(K_1)$ *has index* $r = |nq/mh|$ *in* $\pi_1(U(n))$. *This is the number* r *that occurs in Proposition* 13.2.2.

Proof. We have generators b for $\pi_1(K_1)/\pi_1(K_1')$, $\tilde{b}$ for $\pi_1(\mathbb{C}' \times K_1')/\pi_1(K_1')$, c for $\pi_1(U(n))$ and $\tilde{c}$ for $\pi_1(\mathbb{C}' \times SU(n))$, such that

$$\pi_{\#}(\tilde{b}) = b^{mh}, \quad f_{\#}(b) = c^r, \quad p_{\#}(\tilde{b}) = b^n .$$

The hypothesis says $\tilde{f}_{\#}(\tilde{b}) = (\tilde{c})^q$. Now $c^{rmh} = f_{\#}\pi_{\#}\tilde{b} = p_{\#}\tilde{f}_{\#}\tilde{b} = c^{nq}$.

q.e.d.

Now we go on to consider the parabolics listed in §13.1.

13.3. First consider the real case (13.1.1). There $Z = \mathbb{R} = \mathfrak{z}$ and $\mathfrak{z}^* \cong \mathbb{R}$ under $\lambda(z) = \lambda z$. We consider two possibilities for invariant polynomial,

(13.3.1a) $\psi(\lambda) = \lambda$, degree $d = 1$,

(13.3.1b) $\psi(\lambda) = \lambda^2$, degree $d = 2$.

The corresponding operators on $L^2(P)$ are

(13.3.2a) $D = (\partial/\partial z)^3$, differential but not positive,

(13.3.2b) $D = |\partial/\partial z|^3$, positive but not differential.

In either case, the "unit sphere" $S = \{\lambda \in \mathfrak{z}^*: |\lambda| = 1\} = \{\pm 1\}$, and $M = GL'(2;\mathbb{R})$ acts trivially on it.

13.3.3. Lemma. *Let $0 \neq \lambda \in \mathfrak{z}^*$ and let $[\pi] \in \hat{N}$ be the corresponding unitary representation class. Then $[\pi]$ extends to a class $[\tilde{\pi}] \in (NM)^\wedge$.*

Proof. The representation of M on $\mathfrak{n}/\mathfrak{z}$ is a map $f: GL'(2;\mathbb{R}) \to Sp(2;\mathbb{R})$. On the maximal compact subgroup of the identity component it is

$$f:U(1) \to U(2) \quad \text{by} \quad f(e^{i\theta}) = \begin{pmatrix} e^{3i\theta} & 0 \\ 0 & e^{\pm i\theta} \end{pmatrix} .$$

Factor

$$\begin{pmatrix} e^{3i\theta} & 0 \\ 0 & e^{\pm i\theta} \end{pmatrix} = e^{(2 \text{ or } 1)i\theta} \begin{pmatrix} e^{(1 \text{ or } 2)i\theta} & 0 \\ 0 & e^{-(1 \text{ or } 2)i\theta} \end{pmatrix} .$$

Thus, in Proposition 13.2.6 we have

$$r = |nq/mh| = |2(2 \text{ or } 1)/1|, \quad \text{even},$$

and Proposition 13.2.2 gives vanishing of the Mackey obstruction to extension of $[\pi]$ from N to NM.

q.e.d.

Let $[\pi_{\pm}]$ denote the classes in $\hat{N}$ defined by ± 1 in $\mathfrak{z}^*$, and let $[\tilde{\pi}_{\pm}]$ denote their extensions to NM provided by Lemma 13.3.3. Then the "generic" representations of P are the

$$\pi_{\pm,\nu} = \operatorname{Ind}_{NM\uparrow P}(\tilde{\pi}_{\pm} \otimes \nu), \quad [\nu] \in \hat{M}. \tag{13.3.4}$$

As in [14, §§3 and 4] the Fourier Inversion formula now is

$$F(1_P) = \int_{GL'(2;\mathbb{R})^\wedge} \{\text{trace } \pi_{+,\nu}(DF) + \text{trace } \pi_{-,\nu}(DF)\} d\mu(\nu) \tag{13.3.5}$$

for either choice (13.3.2) of operator D.

13.4. Now consider the complex case (13.1.2). There $Z = \mathbb{C} = \mathfrak{z}$, so $\mathfrak{z}^* \cong \mathbb{C}$ under $\lambda(z) = \mathrm{Re}(\lambda z)$. We use the $\mathbb{R}$-polynomial

$$\psi(\lambda) = |\lambda|^2, \quad \text{degree } d = 2, \tag{13.4.1}$$

which leads to the operator

$$D = \Delta^3, \ \Delta = \text{Laplacian on } Z = \mathbb{R}^2. \tag{13.4.2}$$

$M = GL'(2;\mathbb{C})$ acts transitively on the "unit sphere" $S = \{\lambda \in \mathfrak{z}^* : |\lambda| = 1\}$ by means of its circle factor. The isotropy subgroup at $1 \in S$ is

$$M_1 = SL(2;\mathbb{C}). \tag{13.4.3}$$

The class $[\pi] \in \hat{N}$ for $1 \in \mathfrak{z}^*$ extends to $N \cdot Sp(2;\mathbb{C})$ [28, Theorem 9.19], so it extends to a class $[\tilde{\pi}] \in (NM_1)^\wedge$. Now the "generic" representations of P are the

$$\pi_\nu = \mathrm{Ind}_{NM_1 \uparrow P}(\tilde{\pi} \otimes \nu), \quad \nu \in \widehat{M_1} \tag{13.4.4}$$

and the Fourier Inversion formula is

$$\text{(13.4.5)} \quad \begin{cases} F(1_P) = \displaystyle\int_{SL(2;\mathbb{C})^\wedge} \text{trace } \pi(DF)d\mu(\nu) \\ \text{where } D = \Delta_Z^3 \text{ and } F \text{ is a Schwartz function on } P. \end{cases}$$

§14. Fourier Inversion Inside the Group F_4

We now apply our Fourier Inversion procedure to parabolic subgroups $P = MAN$ in a group G of type F_4: 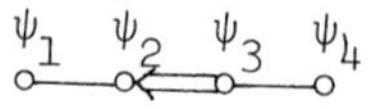

. The parabolics in question derive from $P_{\{\psi_1\}}$ and $P_{\{\psi_4\}}$, and they all satisfy (9.1) because the nilradical is noncommutative. So (6.6.2) and §7.2 give us

14.1.1. Theorem. *The following are the cases in which* $P = MAN$ *is a parabolic subgroup in a simple group* G *of type* F_4, N *has square integrable representations, and (this is automatic)* P *satisfies* (9.1).

1. $G = F_{4,C_3A_1}$ *and* P *is isomorphic either to*

(14.1.2) $$\{\operatorname{Im}\mathfrak{C}_{\mathbb{R}} + \mathfrak{C}_{\mathbb{R}}\}\cdot\{\mathrm{Spin}(3,4) \times \mathbb{R}^+\}$$

or to

(14.1.3) $$\{\mathbb{R} + \mathbb{R}^{14}\}\cdot\{Sp(3;\mathbb{R}) \times \mathbb{R}^+\}\ .$$

2. $G = (F_4)_{\mathbb{C}}$ *and* P *is isomorphic either to*

(14.1.4) $$\{\operatorname{Im}\mathfrak{C}_{\mathbb{C}} + \mathfrak{C}_{\mathbb{C}}\}\cdot\{\mathrm{Spin}(7;\mathbb{C}) \times \mathbb{C}^*\}$$

or to

(14.1.5) $$\{\mathbb{C} + \mathbb{C}^{14}\}\cdot\{Sp(3;\mathbb{C}) \times \mathbb{C}^*\}\ .$$

3. $G = F_{4,B_4}$ <u>and</u> P <u>is isomorphic to</u>

$$\{\mathrm{Im}\,\mathfrak{O} + \mathfrak{O}\}\cdot\{\mathrm{Spin}(7) \times \mathbb{R}^+\} \,. \tag{14.1.6}$$

We now run through these five cases.

<u>14.2</u>. $G = F_{4,C_3A_1}$ <u>and</u> P <u>is given by</u> (14.1.2).

Here $Z = \mathfrak{z} = \mathrm{Im}\,\mathfrak{O}_{\mathbb{R}} \cong \mathbb{R}^{3,4}$, and $\mathfrak{z}^* \cong \mathrm{Im}\,\mathfrak{O}_{\mathbb{R}}$ under $\lambda(z) = \mathrm{Re}\,(\lambda\bar{z})$. $M = \mathrm{Spin}(3,4)$ acts by the vector representation ν: $\overset{1}{\circ}\!-\!\circ\!\Rightarrow\!\circ$. We can use the polynomial invariant

$$\psi(\lambda) = \lambda\bar{\lambda} = \langle\lambda,\lambda\rangle \text{ in } \mathbb{R}^{3,4} \,, \qquad \text{degree } d = 2. \tag{14.2.1}$$

The data k, ℓ, q of (9.3a) are 7, 8, 11; so the positive self-adjoint operator D on $L^2(P)$ is

$$D = |\square|^{11/2} \,, \qquad \square = \text{Laplacian of } \mathbb{R}^{3,4} \,. \tag{14.2.2}$$

The "unit sphere" $S = \{\lambda \in \mathfrak{z}^* : \lambda\bar{\lambda} = \pm 1\}$ decomposes under M as $S_+ \cup S_-$ where

$$\begin{cases} S_+ = \mathrm{Ad}^*(M)\cdot\lambda_+ \cong \mathrm{Spin}(3,4)/\mathrm{Spin}(2,4) \,, \\ S_- = \mathrm{Ad}^*(M)\cdot\lambda_- \cong \mathrm{Spin}(3,4)/\mathrm{Spin}(3,3) \,. \end{cases} \tag{14.2.3}$$

<u>14.2.4. Lemma</u>. <u>Both classes</u> $[\pi_\pm] \in \hat{N}$, <u>defined by</u> $\lambda_\pm \in \mathfrak{z}^*$, <u>have extensions</u> $[\tilde{\pi}_\pm] \in (NM_\pm)^\wedge$.

Proof. Spin $(2,4) \cong SU(2,2)$ with action $\overset{1}{\circ}$—$\circ$—$\circ$ on $\mathfrak{n}/\mathfrak{z}$. Let $N' = N/(\ker \pi_+)^0$, 9-dimensional Heisenberg group. Then $N'\cdot M_+ \cong (\mathrm{Im}\,\mathbb{C} + \mathbb{C}^{2,2})\cdot SU(2,2)$, and [27] π_+ extends from N' to $N'M_+$, hence from N to NM_+.

The maximal compact subgroup of Spin(3,3) is semisimple, hence has finite fundamental group. The remark just after (13.2.3), together with Proposition 13.2.2, gives vanishing of the Mackey obstruction to extension of $[\pi_-]$ from N to NM_-.

q.e.d.

The "generic" representations of P now are the

(14.2.5) $$\pi_{\pm,\nu} = \mathrm{Ind}_{NM_\pm \uparrow P}(\widetilde{\pi_\pm} \otimes \nu), \quad [\nu] \in \widehat{M_\pm},$$

and this leads to a Fourier Inversion formula

(14.2.6) $$\begin{cases} F(1_P) = \displaystyle\int_{\mathrm{Spin}(2,4)^\wedge} \mathrm{trace}\, \pi_{+,\nu}(DF)d\mu_+(\nu) \\ \qquad + \displaystyle\int_{\mathrm{Spin}(3,3)^\wedge} \mathrm{trace}\, \pi_{-,\nu}(DF)d\mu_-(\nu) \end{cases}$$

for all C^∞ functions F on P that satisfy (9.5).

14.3. $G = F_{4,C_3A_1}$ and P is given by (14.1.3).

Here N is the 15-dimensional Heisenberg group, and $M = Sp(3;\mathbb{R})$ acts trivially on $\mathfrak{z}$, acts by $\circ$—$\circ\Leftarrow\overset{1}{\circ}$ on $\mathfrak{n}/\mathfrak{z}$. So the invariant polynomial is $\psi(\lambda) = \lambda$ and the positive self-adjoint operator D

on $L^2(P)$ is

(14.3.1) $$D = \partial^8/\partial z^8, \quad 4^{th} \text{ power of the Laplacian.}$$

The unit sphere $S = \{\pm 1\} \subset \mathfrak{z}^*$. Let $[\pi_\pm] \in \hat{N}$ denote the corresponding classes.

<u>14.3.2. Lemma</u>. <u>Let</u> σ <u>denote the element of order</u> 2 <u>in the circle group</u> $H^2(M;\mathbb{C}') = H^2(Sp(3;\mathbb{R});\mathbb{C}')$. <u>Then</u> $\pi_\pm$ <u>extends to a unitary</u> σ-<u>representation of</u> $N \cdot M$.

<u>Proof</u>. The action of M on $\mathfrak{n}/\mathfrak{z}$ is the representation $f\colon Sp(3;\mathbb{R}) \to Sp(7;\mathbb{R})$ of highest weight $\circ\!-\!\!-\!\circ\!\Leftarrow\!\overset{1}{\circ}$. Using $\Lambda^3(\overset{1}{\circ}\!-\!\!-\!\circ\!\Leftarrow\!\circ) = \overset{1}{\circ}\!-\!\!-\!\circ\!\Leftarrow\!\circ \oplus \circ\!-\!\!-\!\circ\!\Leftarrow\!\overset{1}{\circ}$, an easy calculation, f is given on maximal compact subgroups as $f\colon U(3) \to U(7)$ where, for some $u \in \{0,1,\ldots,6\}$,

$$f(e^{i\theta}I_3) = \begin{pmatrix} e^{3i\theta} & 0 & 0 \\ 0 & e^{i\theta}I_u & 0 \\ 0 & 0 & e^{-i\theta}I_{6-u} \end{pmatrix}$$

$$= e^{(2u-3)i\theta/7} \begin{pmatrix} e^{ai\theta/7} & 0 & 0 \\ 0 & e^{bi\theta/7}I_u & 0 \\ 0 & 0 & e^{ci\theta/7}I_{6-u} \end{pmatrix}$$

where $a = 24-2u$, $b = 10-2u$, $c = -4-2u$. Now, in Proposition 13.2.6 we have $m = n = 7$, $h = 3$, and $q = 2u-3$. It follows that 3 divides u and r is odd, and the assertion follows from Proposition 13.2.2.

<u>q.e.d.</u>

Now the "generic" representations of P are the

$$\pi_{\pm,\nu} = \mathrm{Ind}_{NM\uparrow P}(\tilde{\pi}_{\pm} \otimes \nu), \quad [\nu] \in (M)^{\wedge}_{\sigma} , \tag{14.3.3}$$

which leads to a Fourier Inversion formula

$$\text{(14.3.4)} \quad \begin{cases} F(1_P) = \int_{Sp(3;\mathbb{R})^{\wedge}_{\sigma}} \{\mathrm{trace}\ \pi_{+,\nu}(DF) + \mathrm{trace}\ \pi_{-,\nu}(DF)\} d\mu(\nu) \\ \text{for all Schwartz class functions } F \text{ on } P. \end{cases}$$

<u>14.4.</u> $G = (F_4)_{\mathbb{C}}$ <u>and</u> P <u>is isomorphic to</u> (14.1.4). This is the complexification of the case studied in §14.2. Here $Z = \mathfrak{z} = \mathrm{Im}(\mathfrak{C}_{\mathbb{C}})$, the complement to $1 \cdot \mathbb{C}$ in $\mathfrak{C}_{\mathbb{C}}$ given by $\bar{z} = -z$, and $\mathfrak{z}^* \cong \mathrm{Im}(\mathfrak{C}_{\mathbb{C}}) \cong \mathbb{R}^{7,7}$ under $\lambda(z) = \frac{1}{2}\{\lambda\bar{z} + z\bar{\lambda}\} = \langle \lambda, z \rangle$. The M-invariant $\mathbb{R}$-polynomial is

$$\psi(\lambda) = |\lambda\bar{\lambda}|^2, \quad \text{degree } d = 4. \tag{14.4.1}$$

The data of (9.3a) is 14, 16, 22. In terms of

$$\Box \cong \sum_1^7 \partial^2/\partial z_j^2 - \sum_8^{14} \partial^2/\partial z_j^2 , \quad \text{Laplacian on } \mathrm{Im}\mathfrak{C}_{\mathbb{C}} \cong \mathbb{R}^{7,7}$$

the self-adjoint operator D on $L^2(P)$ is either of

(14.4.2a) $D = |\Box|^{11}$, positive but not differential,

(14.4.2b) $D = \Box^{11}$, differential but not positive.

The "unit sphere" $S = \{\lambda \in \mathfrak{z}^*: |\lambda\bar{\lambda}| = 1\}$ is a single orbit $Ad^*(M)\cdot\lambda_1 \cong M/M_1$. Here

(14.4.3) $M = Spin(7;\mathbb{C}) \times \mathbb{C}'$ and $M_1 \cong Spin(6;\mathbb{C}) \times \{\pm 1, \pm\sqrt{-1}\}$.

Since $Spin(6;\mathbb{C})$ is a simply connected semisimple group, Mackey obstructions vanish, and the class $[\pi] \in \hat{N}$ for $\lambda_1 \in \mathfrak{z}^*$ extends to a class $[\tilde{\pi}] \in (NM_1)^{\wedge}$. So we have "generic" representations

(14.4.4) $\pi_\nu = Ind_{NM_1 \uparrow P}(\tilde{\pi} \otimes \nu), \quad [\nu] \in \widehat{M_1},$

and a Fourier Inversion formula

(14.4.5) $F(1_P) = \int_{\{Spin(6;\mathbb{C})\times\mathbb{Z}_4\}^{\wedge}} \text{trace } \pi_\nu(DF)d\mu(\nu)$

with $F \in C^\infty(P)$ constrained according to the choice of (14.4.2).

<u>14.5</u>. $G = (F_4)_{\mathbb{C}}$ <u>and</u> P <u>is given by</u> (14.1.5), complexification of the case examined in §14.3.

Here $Z = \mathfrak{z} = \mathbb{C}$ and $\mathfrak{z}^* \cong \mathbb{C}$ under $\lambda(z) = Re(\lambda\bar{z})$. The invariant $\mathbb{R}$-polynomial is $\psi(\lambda) = |\lambda|^2$, and the positive self-adjoint operator is

(14.5.1) $D = \Delta^8$, 8-th power of the Laplacian of $Z = R^2$.

$M = Sp(3;\mathbb{C}) \times \mathbb{C}'$ is transitive on the "unit sphere" $S = \{\lambda \in \mathfrak{z}^*: |\lambda| = 1\}$, with isotropy subgroup $M_1 - Sp(3;\mathbb{C}) \times \{\pm 1\}$

at $\lambda_1 = 1$. The class $[\pi] \in \hat{N}$ for $\lambda_1 \in \mathfrak{z}^*$ extends to a class $[\tilde{\pi}]$ in $(NM_1)^\wedge$ since $Sp(3;\mathbb{C})$ is a simply connected semisimple group. Thus we have "generic" representations

$$\pi_\nu = \mathrm{Ind}_{NM_1 \uparrow P}(\tilde{\pi} \otimes \nu), \quad [\nu] \in \widehat{M_1} \tag{14.5.2}$$

and a Fourier Inversion formula

$$(14.5.3) \qquad \begin{cases} F(1_P) = \displaystyle\int_{\{Sp(3;\mathbb{C}) \times \mathbf{Z}_2\}^\wedge} \text{trace } \pi_\nu(DF) d\mu(\nu) \\ \\ \text{for all Schwartz class functions } F: P \to \mathbb{C} \end{cases} .$$

<u>14.6.</u> $G = F_{4,B_4}$ <u>and</u> P <u>is given by</u> (14.1.6). This is the F_4 case of [9].

Here $Z = \mathfrak{z} = \mathrm{Im}\,\mathfrak{C} \cong \mathbb{R}^7$, and $\mathfrak{z}^* \cong \mathrm{Im}\,\mathfrak{C} \cong \mathbb{R}^7$ under $\lambda(z) = -\mathrm{Re}(\lambda\bar{z}) = \langle \lambda, z \rangle$. $M = Spin(7)$ acts on $\mathfrak{z}^*$ by the vector representation, so we have the polynomial invariant

$$\psi(\lambda) = \lambda\bar{\lambda} = \|\lambda\|^2 , \quad \text{degree } d = 2. \tag{14.6.1}$$

That leads to the positive self-adjoint operator D on $L^2(P)$ given by

$$D = \Delta^{11/2}, \qquad \Delta = \text{Laplacian of } \mathbb{R}^7 . \tag{14.6.2}$$

The unit sphere $S = \{\lambda \in \mathfrak{z}^* : \|\lambda\| = 1\}$ is a single orbit $Ad^*(M)\cdot\lambda_1 \cong Spin(7)/Spin(6)$, and the class $[\pi] \in \hat{N}$ for $\lambda_1 \in \mathfrak{z}^*$

extends to $[\tilde{\pi}] \in (NM_1)^\wedge$ because $M_1 = \text{Spin}(6)$ is simply connected and semisimple. So we have "generic" representations

$$\pi_\nu = \text{Ind}_{NM_1 \uparrow P}(\tilde{\pi} \otimes \nu), \quad [\nu] \in \widehat{M_1}. \tag{14.6.3}$$

Now we have the usual inversion formula for an appropriate normalization $\mu(\nu) = c \cdot \dim(\nu)$ of Plancherel measure on $M_1 = \text{Spin}(6)^\wedge$. See [9 , (2.5)] for a convenient determination of c. The Fourier Inversion formula is

$$(14.6.4) \quad \begin{cases} F(1_P) = c \sum_{[\nu] \in \text{Spin}(6)^\wedge} \text{trace } \pi_\nu(D) \cdot \dim(\nu) \\ \\ \text{for all Schwartz class functions } F \text{ on } P . \end{cases}$$

§15. Fourier Inversion Inside the Group E_6

We apply the Fourier Inversion method to parabolic subgroups $P = MAN$ in a group G of type E_6: 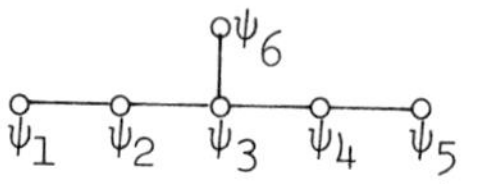

. The parabolics in question derive from $P_{\{\psi_1\}} \cong P_{\{\psi_5\}}$, $P_{\{\psi_6\}}$ and $P_{\{\psi_1,\psi_5\}}$. Since the exceptional 16-dimensional bounded symmetric domain $E_{6,D_5T_1}/U(1)\cdot Spin(10)$ is not of tube type, Theorem 9.15 says that the first of these three classes does not satisfy (9.1). So (6.6.3) and §7.3 give

15.1.1. Theorem. The following are the cases in which $P = MAN$ is a parabolic subgroup in a simple group G of type E_6, N has square integrable representations, and P satisfies (9.1).

1. $G = E_{6,C_4}$ and P is isomorphic either to

(15.1.2) $$\{\mathbb{R} + \mathbb{R}^{20}\}\cdot GL(6;\mathbb{R})$$

or to

(15.1.3) $$\left\{ \begin{pmatrix} a & x & z \\ 0 & b & y \\ 0 & 0 & c \end{pmatrix} : \begin{array}{l} x,\ y,\ z \in \mathfrak{C}_{\mathbb{R}} \\ a,\ b,\ c \in \mathbb{R}^+ \\ abc = 1 \end{array} \right\}\cdot Spin(4,4)$$

Here recall that $\mathfrak{C}_{\mathbb{R}}$ is the split (has zero-divisors) real Cayley algebra.

2. $G = (E_6)_{\mathbb{C}}$ *and* P *is isomorphic either to*

(15.1.4) $$\{\mathbb{C} + \mathbb{C}^{20}\}\cdot\{GL(6;\mathbb{C})/\mathbb{Z}_3\}$$

or to

(15.1.5) $$\left\{\begin{pmatrix} a & x & z \\ 0 & b & y \\ 0 & 0 & c \end{pmatrix} : \begin{array}{l} x, y, z \in \mathfrak{C}_{\mathbb{C}} \\ a, b, c \in \mathbb{C}^* \\ abc = 1 \end{array}\right\}\cdot \mathrm{Spin}(8;\mathbb{C}).$$

Here recall that $\mathfrak{C}_{\mathbb{C}}$ *is the Cayley algebra over the complex numbers.*

3. $G = E_{6,A_1A_5}$ *and* P *is isomorphic either to*

(15.1.6) $$\{\mathbb{R} + \mathbb{R}^{20}\}\cdot\{(SU(3,3)/\mathbb{Z}_3) \times \mathbb{R}^+\}$$

or to

(15.1.7) $$\left\{\begin{pmatrix} a & x & z \\ 0 & b & y \\ 0 & 0 & c \end{pmatrix} : \begin{array}{l} x, y, z \in \mathfrak{C}_{\mathbb{R}} \\ a, b, c \in \mathbb{R}^+ \\ abc = 1 \end{array}\right\}\cdot \mathrm{Spin}(3,5).$$

4. $G = E_{6,D_5T_1}$ *and* P *is isomorphic either to*

(15.1.8) $$\{\mathbb{R} + \mathbb{R}^{20}\}\cdot\{(SU(1,5)/\mathbb{Z}_3) \times \mathbb{R}^+\}$$

or to

$$(15.1.9)\qquad \left\{\begin{pmatrix} a & x & z \\ 0 & b & y \\ 0 & 0 & c \end{pmatrix} : \begin{array}{l} x, y, z \in \mathfrak{C} \\ a, b, c \in \mathbb{R}^+ \\ abc = 1 \end{array}\right\} \cdot \mathrm{Spin}(1,7).$$

Here recall that $\mathfrak{C}$ is the real Cayley division algebra.

5. $G = E_{6,F_4}$ and P is isomorphic to

$$(15.1.10)\qquad \left\{\begin{pmatrix} a & x & z \\ 0 & b & y \\ 0 & 0 & c \end{pmatrix} : \begin{array}{l} x, y, z \in \mathfrak{C} \\ a, b, c \in \mathbb{R}^+ \\ abc = 1 \end{array}\right\} \cdot \mathrm{Spin}(8).$$

We now run through these nine cases.

15.2. $G = E_{6,C_4}$ and P is given by (15.1.2).

Here $Z = \mathfrak{z} = \mathbb{R} \cong \mathfrak{z}^*$ under $\lambda(z) = \lambda z$, and $M = GL'(6;\mathbb{R})$ acts trivially on $\mathfrak{z}^*$. Take $\psi(\lambda) = \lambda$ as invariant polynomial. The data k, ℓ, q of (9.3a) are 1, 20, 11, so we have a choice of self-adjoint operator on $L^2(P)$,

$$(15.2.1a)\qquad D = \partial^{11}/\partial z^{11}, \quad \text{differential but not positive}$$

$$(15.2.1b)\qquad D = |\partial/\partial z|^{11}, \quad \text{positive but not differential.}$$

The "unit sphere" $S = \{\lambda \in \mathfrak{z}^*: |\psi(\lambda)| = 1\} = \{\pm 1\}$. Let $[\pi_+]$, $[\pi_-] \in \hat{N}$ denote the corresponding unitary representation classes.

15.2.2. Lemma. Both $[\pi_\pm] \in \hat{N}$ have extensions $[\widetilde{\pi_\pm}] \in (NM)^\wedge$.

Proof. The action of M on $\mathfrak{n}/\mathfrak{z}$ is the representation $f\colon GL'(6;\mathbb{R}) \to Sp(10;\mathbb{R})$ given by $\circ\!-\!\circ\!-\!\overset{1}{\circ}\!-\!\circ\!-\!\circ$. The maximal compact subgroup $O(6)$ of M has $\pi_1(O(6))$ finite, so $f_\#\colon \pi_1(M) \to \pi_1(Sp(10;\mathbb{R}))$ is trivial and the result follows from Proposition 13.2.2.

q.e.d.

The "generic" representations of P now are the

$$(15.2.3) \qquad \pi_{\pm,\nu} = \mathrm{Ind}_{NM\uparrow P}(\widetilde{\pi_\pm} \otimes \nu), \qquad [\nu] \in \hat{M}.$$

That leads to the Fourier Inversion formula

$$(15.2.4) \qquad F(1_P) = \int_{GL'(6;\mathbb{R})^\wedge} \{\mathrm{trace}\ \pi_{+,\nu}(DF) + \mathrm{trace}\ \pi_{-,\nu}(DF)\} d\mu(\nu)$$

for Schwartz class functions $F\colon P \to \mathbb{C}$. Here [9 , Theorem 2.7] shows that the choice (15.2.1) imposes no restriction.

15.3. $G = E_{6,C_4}$ and P is given by (15.1.3).

Here $Z = \mathfrak{z} = \mathfrak{C}_{\mathbb{R}}$ and $\mathfrak{z}^* \cong \mathfrak{C}_{\mathbb{R}}$ under $\lambda(z) = \mathrm{trace}(\lambda\bar{z})$ using the trace of $\mathfrak{C}_{\mathbb{R}}$. $M = Spin(4,4)$ acts on $\mathfrak{z}^*$ by the vector representation 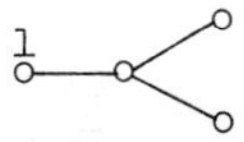 , thus has polynomial invariant

$$(15.3.1) \qquad \psi(\lambda) = \langle \lambda, \lambda \rangle \ \text{ in } \mathfrak{C}_{\mathbb{R}} \cong \mathbb{R}^{4,4}, \quad \text{degree } d = 2.$$

The data k, ℓ, q of (9.3a) are 8, 16, 16; so the positive self-adjoint operator D on $L^2(P)$ is

(15.3.2) $D = \Box^8, \quad \Box = \text{Laplacian of } \mathbb{R}^{4,4}.$

The "unit sphere" $S = \{\lambda \in \mathfrak{z}^* : \langle\lambda,\lambda\rangle = \pm 1\}$ decomposes under M as $S_+ \cup S_-$ where

(15.3.3) $S_\pm = Ad^*(M)\cdot\lambda_\pm \cong Spin(4,4)/Spin(3,4).$

Here $A_1 \cong \mathbb{R}^+$, given by $a, b, c = a, a^{-2}, a$ in (15.1.3).

15.3.4. Lemma. Both classes $[\pi_\pm] \in \hat{N}$ extend to classes $[\widetilde{\pi_\pm}] \in (NA_1M_\pm)^\wedge$.

Proof. Realize $\pi = \pi_\pm$ as $Ind_{Q\uparrow N}(\chi)$ where

$$Q = \left\{ \begin{pmatrix} 1 & 0 & z \\ 0 & 1 & y \\ 0 & 0 & 1 \end{pmatrix} : y, z \in \mathfrak{C}_{\mathbb{R}} \right\} \text{ and where } \chi\begin{pmatrix} 1 & 0 & z \\ 0 & 1 & y \\ 0 & 0 & 1 \end{pmatrix} = e^{\sqrt{-1}\lambda(z)},$$

$\lambda = \lambda_\pm$. Now χ extends to a unitary character $\tilde{\chi}$ on $QA_1M_\pm$ by the formula

$$\tilde{\chi}\left(\begin{pmatrix} a & 0 & z \\ 0 & a^{-2} & y \\ 0 & 0 & a \end{pmatrix}, m\right) = e^{\sqrt{-1}\lambda(z)/a}$$

since m fixes λ. The desired extension is

$$\tilde{\pi}_\pm = Ind_{QA_1M_\pm\uparrow NA_1M_\pm}(\tilde{\chi}). \qquad \text{q.e.d.}$$

Now the "generic" representations of P are the

$$(15.3.5) \qquad \pi_{\pm,\nu} = \mathrm{Ind}_{NA_1M_\pm \uparrow P}(\widetilde{\pi_\pm} \otimes \nu), \quad [\nu] \in (A_1M_\pm)^\wedge .$$

That leads to the Fourier Inversion formula

$$(15.3.6) \qquad F(1_P) = \int_{\{\mathrm{Spin}(3,4)\times\mathbb{R}^+\}^\wedge} \{\mathrm{trace}\ \pi_{+,\nu}(DF) + \mathrm{trace}\ \pi_{-,\nu}(DF)\} d\mu(\nu).$$

<u>15.4</u>. $G = (E_6)_{\mathbb{C}}$ <u>and</u> P <u>is given by</u> (15.1.4).

Here $Z = \mathfrak{z} = \mathbb{C}$ and $\mathfrak{z}^* \cong \mathbb{C}$ under $\lambda(z) = \mathrm{Re}(\lambda\bar{z})$, and $M = GL'(6;\mathbb{C})/\mathbb{Z}_3$ acts on $\mathfrak{z}^*$ by

$$(15.4.1) \qquad \mathrm{Ad}^*(g)\lambda = \det_{\mathbb{C}}(g)^{-6}\cdot\lambda \qquad \text{for } g \in GL'(6;\mathbb{C}),$$

since M acts on $\mathfrak{n}/\mathfrak{z}$ by $\Lambda^3(g)$. Our invariant $\mathbb{R}$-polynomial is

$$(15.4.2) \qquad \psi(\lambda) = |\lambda|^2 , \qquad \text{degree } d = 2.$$

The data k, ℓ, q of (9.3a) are 2, 40, 22; so our invertible positive self-adjoint operator D on $L^2(P)$ is

$$(15.4.3) \qquad D = \Delta^{11} , \quad \Delta = \text{Laplacian on } Z \cong \mathbb{R}^2.$$

The "unit sphere" $S = \{\lambda \in \mathfrak{z}^* : |\lambda| = 1\}$ is an orbit of the center of M, and from (15.4.1) we see that the isotropy subgroup of M at any $\lambda \in S$ is

(15.4.4) $M_1 = \{g \in GL(6;\mathbb{C}) : (\det_{\mathbb{C}} g)^6 = 1\}/\mathbb{Z}_3.$

Let $[\pi] \in \hat{N}$ correspond to $1 \in S$. Then [28, Theorem 9.19] implies that $[\pi]$ extends to a class $[\tilde{\pi}] \in (NM_1)\hat{}$. So the "generic" representations of P are the

(15.4.5) $\pi_\nu = \mathrm{Ind}_{NM_1 \uparrow P}(\tilde{\pi} \otimes \nu), \quad [\nu] \in \widehat{M_1}.$

The corresponding Fourier Inversion formula is

(15.4.6) $F(1_P) = \int_{\widehat{M_1}} \mathrm{trace}\ \pi_\nu(\Delta^{11} F)\, d\mu(\nu)$

for Schwartz class functions $F : P \to \mathbb{C}$.

<u>15.5.</u> $G = (E_6)_{\mathbb{C}}$ <u>and</u> P <u>is given by</u> (15.1.5).

Here $Z = \mathfrak{z} = \mathfrak{C}_{\mathbb{C}}$ and $\mathfrak{z}^* \cong \mathfrak{C}_{\mathbb{C}}$ under $\lambda(z) = \mathrm{Re\ Trace}(\lambda\bar{z})$ using the trace of $\mathfrak{C}_{\mathbb{C}}$. $M = \mathrm{Spin}(8;\mathbb{C}) \times \mathbb{C}' \times \mathbb{C}'$ acts on $\mathfrak{z}^*$ by

(15.5.1) $\mathrm{Ad}^*\left(g, \begin{pmatrix} a & 0 & 0 \\ 0 & b & 0 \\ 0 & 0 & c \end{pmatrix}\right) : \lambda \mapsto a^{-1} c \cdot \nu(g)\lambda$, ν: 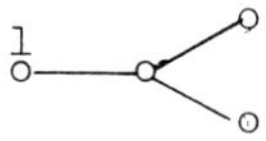 ,

where $abc = 1$ as in (15.1.5) and where $|a|, |b|, |c| = 1$. This gives us the $\mathbb{R}$-polynomial invariant

(15.5.2) $\psi(\lambda) = |\langle \lambda, \lambda \rangle|^2$, degree $d = 4$,

where $\langle\ ,\ \rangle$ is the $\mathbb{C}$-bilinear invariant of 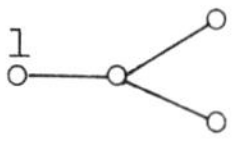 . The data k, ℓ, q of (9.3a) are 16, 32, 32. So the positive invertible self-adjoint operator D on $L^2(P)$ is

$$(15.5.3) \qquad D = \Psi^8, \text{ always differential.}$$

The "unit sphere" $S = \{\lambda \in \mathfrak{z}^* : |\langle\lambda,\lambda\rangle| = 1\}$ is, from (15.5.1), a single orbit of M, which acts with isotropy subgroup

$$(15.5.4) \qquad M_1 \cong \mathrm{Spin}(7;\mathbb{C}) \times \mathbb{C}'.$$

Here $A \cong \mathbb{R}^+ \times \mathbb{R}^+$ with $A_1 = \left\{ \begin{pmatrix} a & & \\ & a^{-2} & \\ & & a \end{pmatrix} : a \in \mathbb{R}^+ \right\}$, so

$$(15.5.5) \qquad A_1M_1 \cong \mathrm{Spin}(7;\mathbb{C}) \times \mathbb{C}^*.$$

Let $[\pi] \in \hat{N}$ correspond to the $\lambda \in \mathfrak{z}^*$ fixed by a choice of A_1M_1. As in Lemma 15.3.4, one concludes that $[\pi]$ extends to a class $[\tilde{\pi}] \in (NA_1M_1)^{\wedge}$. So the "generic" representations of P are the

$$(15.5.6) \qquad \pi_\nu = \mathrm{Ind}_{NA_1M_1 \uparrow P}(\tilde{\pi} \otimes \nu), \ [\nu] \in (A_1M_1)^{\wedge}.$$

This leads to the Fourier Inversion formula

$$(15.5.7) \qquad F(1_P) = \int_{\{\mathrm{Spin}(7;\mathbb{C}) \times \mathbb{C}^*\}^{\wedge}} \operatorname{trace} \pi_\nu(DF)\,d\mu(\nu)$$

for Schwartz class functions F on P.

15.6. $G = E_{6,A_1A_5}$ and P is given by (15.1.6).

Here $Z = \mathfrak{z} = \mathbb{R}$ and $\mathfrak{z}^* \cong \mathbb{R}$ under $\lambda(z) = \lambda z$, and $M = SU(3,3)/\mathbb{Z}_3$ acts trivially on $\mathfrak{z}^*$. Take $\psi(\lambda) = \lambda$ as invariant polynomial. The data k, ℓ, q of (9.3a) are 1, 20, 11; so we have a choice of invertible self-adjoint operator D on $L^2(P)$,

(15.6.1a) $\qquad D = (\partial/\partial z)^{11}$, differential but not positive,

(15.6.1b) $\qquad D = |\partial/\partial z|^{11}$, positive but not differential.

The "unit sphere" $S = \{\lambda \in \mathfrak{z}^* : |\lambda| = 1\} = \{\pm 1\}$. Let $[\pi_+], [\pi_-] \in \hat{N}$ denote the corresponding equivalence classes of unitary representations.

15.6.2. Lemma. Both $[\pi_\pm] \in \hat{N}$ extend to classes $[\tilde{\pi}_\pm] \in (NM)\hat{}$.

Proof. M acts on $\mathfrak{n}/\mathfrak{z}$ by $\circ\!-\!\circ\!-\!\overset{1}{\circ}\!-\!\circ\!-\!\circ$, which we write as $f: SU(3,3) \to Sp(10;\mathbb{R})$. The map on maximal compact subgroups is $f: S(U(3) \times U(3)) \to U(10)$, and it acts on the center of the maximal compact subgroup by

$$f\begin{pmatrix} e^{i\theta}I_3 & 0 \\ 0 & e^{-i\theta}I_3 \end{pmatrix} = \begin{pmatrix} e^{3i\theta} & 0 & 0 \\ 0 & e^{i\theta}I_u & 0 \\ 0 & 0 & e^{-i\theta}I_{9-u} \end{pmatrix}$$

$$= e^{(u-3)i\theta/5}\begin{pmatrix} e^{(18-u)i\theta/5} & 0 & 0 \\ 0 & e^{(8-u)i\theta/5}I_u & 0 \\ 0 & 0 & e^{-(u+2)i\theta/5}I_{9-u} \end{pmatrix}$$

for some integer $u \in \{0,1,\dots,9\}$. So, in Proposition 13.2.6 we have $n = 10$, $m = 5$, $q = u-3$ and $h = 3$, thus $r = |2(u-3)/3|$, so 3 divides u and r is even in Proposition 13.2.2. We conclude that the Mackey obstruction to extension of $[\pi_\pm]$ must vanish.

<u>q.e.d.</u>

The "generic" representations of P now are the

$$\pi_{\pm,\nu} = \mathrm{Ind}_{NM\uparrow P}(\widetilde{\pi_\pm} \otimes \nu), \quad [\nu] \in \hat{M} \quad . \tag{15.6.3}$$

As before, that leads to a Fourier Inversion formula

$$\begin{cases} F(1_P) = \displaystyle\int_{\{SU(3,3)/\mathbb{Z}_3\}^\wedge} \{\mathrm{trace}\ \pi_{+,\nu}(DF) + \mathrm{trace}\ \pi_{-,\nu}(DF)\}d\mu(\nu) \\ \text{for all Schwartz class functions } F \text{ on } P. \end{cases} \tag{15.6.4}$$

<u>15.7</u>. $G = E_{6,A_1A_5}$ <u>and</u> P <u>is given by</u> (15.1.7).

Here $Z = \mathfrak{C}_{\mathbb{R}} = \mathfrak{z}$ and $\mathfrak{z}^* \cong \mathfrak{C}_{\mathbb{R}}$ under $\lambda(z) = \mathrm{trace}\ (\lambda\bar{z})$ using the trace in $\mathfrak{C}_{\mathbb{R}}$. $M = Spin(3,5)$ acts on $\mathfrak{z}$, hence (the representation is self-dual) on $\mathfrak{z}^*$ by the vector representation 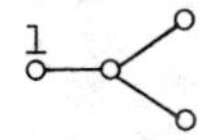 . The image of M in this action is $SO(3,5)$ and the polynomial invariant is

$$\psi(\lambda) = \langle\lambda,\lambda\rangle \quad \text{on } \mathfrak{C}_{\mathbb{R}} \cong \mathbb{R}^{3,5}, \quad \text{degree } d = 2. \tag{15.7.1}$$

The data k , ℓ, q of (9.3a) are 8, 16, 16; so the invertible positive self-adjoint operator D on $L^2(P)$ is

$$(15.7.2) \qquad D = \square^8, \quad \square = \text{Laplacian of } \mathbb{R}^{3,5} .$$

The "unit sphere" $S = \{\lambda \in \mathfrak{z}^* : \langle \lambda,\lambda \rangle = \pm 1\}$ decomposes as the union of two M-orbits,

$$(15.7.3) \qquad \begin{cases} S_+ = \mathrm{Ad}^*(M)\cdot\lambda_+ \cong \mathrm{Spin}(3,5)/\mathrm{Spin}(2,5) , \\ S_- = \mathrm{Ad}^*(M)\cdot\lambda_- \cong \mathrm{Spin}(3,5)/\mathrm{Spin}(3,4) . \end{cases}$$

Here $A_1 \cong \mathbb{R}^+$, given by $a, b, c = a, a^{-2}, a$ in (15.1.7). Let $[\pi_\pm] \in \hat{N}$ be the classes defined by $\lambda_\pm \in \mathfrak{z}^*$. Arguing as in Lemma 15.3.4 we extend $[\pi_\pm]$ to classes $[\widetilde{\pi_\pm}] \in (NA_1M_\pm)^\wedge$, so the "generic" representations of P are just the

$$(15.7.4) \qquad \pi_{\pm,\nu} = \mathrm{Ind}_{NA_1M_\pm \uparrow P}(\widetilde{\pi_\pm} \otimes \nu), \quad [\nu] \in (A_1M_\pm)^\wedge ,$$

and this leads to a Fourier Inversion formula

$$(15.7.5) \qquad \begin{cases} F(1_P) = \displaystyle\int_{\{\mathrm{Spin}(2,5) \times \mathbb{R}^+\}^\wedge} \text{trace } \pi_{+,\nu}(DF)d\mu_+(\nu) \\ \qquad + \displaystyle\int_{\{\mathrm{Spin}(3,4) \times \mathbb{R}^+\}^\wedge} \text{trace } \pi_{-,\nu}(DF)d\mu_-(\nu) \end{cases}$$

for Schwartz class functions $F: P \to \mathbb{C}$.

<u>15.8.</u> $G = E_{6,D_5T_1}$ <u>and</u> P <u>is given by</u> (15.1.8).

Here $Z = \mathbb{R} = \mathfrak{z}$ with $\mathfrak{z}^* \cong \mathbb{R}$ under $\lambda(z) = \lambda z$, and $M = SU(1,5)/\mathbb{Z}_3$ acts trivially on $\mathfrak{z}^*$. So $\psi(\lambda) = \lambda$ is invariant, $q = 11$ in (9.3a), and our self adjoint operator on $L^2(P)$ is either one of

$$D = \partial^{11}/\partial z^{11}, \quad \text{differential but not positive} \tag{15.8.1a}$$

$$D = |\partial/\partial z|^{11}, \quad \text{positive but not differential.} \tag{15.8.1b}$$

The "unit sphere" $S = \{\lambda \in \mathfrak{z}^* : |\psi(\lambda)| = 1\} = \{\pm 1\}$. Let $[\pi_\pm] \in \hat{N}$ denote the corresponding unitary representation classes.

15.8.2. Lemma. *The $[\pi_\pm] \in \hat{N}$ extend to unitary representation classes $[\tilde{\pi}_\pm] \in (NM)^\wedge$.*

Proof. The action of M on $\mathfrak{n}/\mathfrak{z}$ is the representation $f\colon SU(1,5) \to Sp(10;\mathbb{R})$ given by $\overset{}{\circ}\!-\!\circ\!-\!\overset{1}{\circ}\!-\!\circ\!-\!\circ$. Restrict f to the maximal compact subgroup

$$S(U(1) \times U(5)) = \left\{ \begin{pmatrix} \det A^{-1} & 0 \\ 0 & A \end{pmatrix} : A \in U(5) \right\} \cong U(5).$$

There it is given on the center by

$$f(e^{i\theta} I_5) = \begin{pmatrix} e^{3i\theta} I_u & 0 \\ 0 & e^{-3i\theta} I_{10-u} \end{pmatrix}$$

$$= e^{(3u-15)i\theta/5} \begin{pmatrix} e^{(30-3u)i\theta/5} I_u & 0 \\ 0 & e^{-3ui\theta/5} I_{10-u} \end{pmatrix}$$

for some integer $u \in \{0,1,\ldots,10\}$. In Proposition 13.2.6, now, $n = 10$, $m = 5$, $h = 5$ and $q = 3u-15$, so $r = |2(3u-15)/5|$. Thus 5 divides u and r is even in Proposition 13.2.2; that kills the Mackey obstruction.

<u>q.e.d.</u>

The "generic" representations of P now are the

$$(15.8.3) \qquad \pi_{\pm,\nu} = \mathrm{Ind}_{NM\uparrow P}(\widetilde{\pi_{\pm}} \otimes \nu), \qquad [\nu] \in \hat{M}.$$

That leads to the Fourier Inversion formula

$$(15.8.4) \qquad F(1_P) = \int_{\{SU(1,5)/\mathbb{Z}_3\}^{\wedge}} \{\mathrm{trace}\ \pi_{+,\nu}(DF) + \mathrm{trace}\ \pi_{-,\nu}(DF)\} d\mu(\nu)$$

where F is a Schwartz class function on P.

<u>15.9</u>. $G = E_{6,D_5T_1}$ <u>and</u> P <u>is given by</u> (15.1.9).

Here $Z = \mathfrak{C} = \mathfrak{z}$ with $\mathfrak{z}^* \cong \mathfrak{C}$ under $\lambda(z) = \mathrm{trace}\ (\lambda\bar{z})$, using trace in the Cayley division algebra $\mathfrak{C}$. $M = Spin(1,7)$ acts on $\mathfrak{z}^*$ by the vector representation 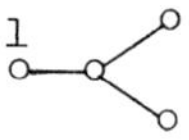 , through $SO(1,7)$, which has polynomial invariant

$$(15.9.1) \qquad \psi(\lambda) = \langle \lambda,\lambda \rangle \quad \text{in } \mathfrak{C} \cong \mathbb{R}^{1,7}, \qquad \text{degree } d = 2.$$

Now the positive self adjoint operator on $L^2(P)$ is

$$(15.9.2) \qquad D = \Box^8, \quad \Box = \text{Laplacian of } \mathbb{R}^{1,7} .$$

The "unit sphere" $S = \{\lambda \in \mathfrak{z}^*: \langle\lambda,\lambda\rangle = \pm 1\}$ decomposes under M into two orbits

$$(15.9.3)\qquad \begin{cases} S_+ = Ad^*(M)\cdot\lambda_+ \cong Spin(1,7)/Spin(7)\ , \\ S_- = Ad^*(M)\cdot\lambda_- \cong Spin(1,7)/Spin(1,6)\ . \end{cases}$$

As before, $A_1 \cong \mathbb{R}^+$ given a, b, c = a, a^{-2}, a in (15.1.9), and the argument of Lemma 15.3.4 extends

$$[\pi_\pm]: \text{ class in } \hat{N} \text{ defined by } \lambda_\pm \in \mathfrak{z}^*$$

from N to $NA_1M_\pm$. So P has "generic" representations

$$(15.9.4)\qquad \pi_{\pm,\nu} = Ind_{NA_1M_\pm\uparrow P}(\widetilde{\pi}_\pm \otimes \nu), \qquad [\nu] \in (A_1M_\pm)^\wedge\ .$$

Thus we come to the Fourier Inversion formula

$$(15.9.5)\qquad \begin{cases} F(1_P) = \displaystyle\int_{\{Spin(7)\times\mathbb{R}^+\}^\wedge} \text{trace}\ \pi_{+,\nu}(DF)d\mu_+(\nu) \\ \qquad\qquad + \displaystyle\int_{\{Spin(1,6)\times\mathbb{R}^+\}^\wedge} \text{trace}\ \pi_{-,\nu}(DF)d\mu_-(\nu) \end{cases}$$

for Schwartz class functions $F: P \to \mathbb{C}$.

15.10. $G = E_{6,F_4}$ and P is given by (15.1.10).
This is the $SL(3,\mathfrak{C})$ case of [14, Proposition 5.20]. $Z = \mathfrak{C} = \mathfrak{z}$, $\mathfrak{z}^* \cong \mathfrak{C}$ as before, $\psi(\lambda) = \langle\lambda,\lambda\rangle$ on $\mathfrak{C} \cong \mathbb{R}^8$, and

(15.10.1) $$D = \Delta^8, \quad \Delta = \text{Laplacian on } \mathbb{R}^8.$$

The unit sphere $S = \{\lambda \in \mathfrak{z}^* : \langle \lambda, \lambda \rangle = 1\}$, genuine sphere, is $Ad^*(M)\cdot\lambda_0 \cong Spin(8)/Spin(7)$. As above, the inversion formula is

(15.10.2) $$F(1_P) = \int_{\{Spin(7) \times \mathbb{R}^+\}^\wedge} \text{trace } \pi_\nu(DF)d\mu(\nu)$$

for Schwartz class functions F on P.

§16. Fourier Inversion Inside the Group E_7

We now apply our Fourier Inversion procedure to parabolic subgroups $P = MAN$ in a group G of type

E_7: 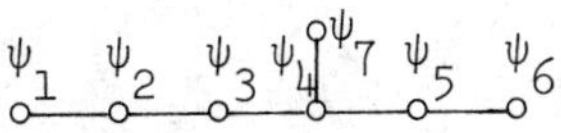. The parabolics in question derive from

$P_{\{\psi_1\}}$, $P_{\{\psi_2\}}$ and $P_{\{\psi_6\}}$.

16.1.1. The exceptional 27-dimensional bounded symmetric domain $E_{7,E_6T_1}/E_6T_1$ is of tube type. Theorem 9.15 says that the parabolics that derive from $P_{\{\psi_1\}}$ satisfy (9.1). The proof of Theorem 9.15 provides a polynomial invariant of degree 3 there, which is the norm or Freudenthal determinant for an exceptional simple Jordan algebra structure on the nilradical.

Now (6.6.4) and §7.4 give

16.1.2. Theorem. *The following are the cases in which* $P = MAN$ *is a parabolic subgroup in a simple group* G *of type* E_7, N *has square integrable representations, and (this is automatic)* P *satisfies* (9.1).

1. $G = E_{7,A_7}$ *and* P *is isomorphic either to*

(16.1.3) $\mathbb{R}^{27}\cdot\{E_{6,C_4} \times \mathbb{R}^+\}$

or to

(16.1.4) $\{\mathbb{R}^{5,5} + (\mathbb{R}^2 \otimes \mathbb{R}^{16})\}\cdot\{SL(2;\mathbb{R}) \times Spin(5,5) \times \mathbb{R}^+\}$

or to

(16.1.5) $\{\mathbb{R} + \mathbb{R}^{32}\}\cdot\{Spin(6,6) \times \mathbb{R}^{+}\}$.

2. $G = (E_7)_{\mathbb{C}}$ <u>and</u> P <u>is isomorphic either to</u>

(16.1.6) $\mathbb{C}^{27}\cdot\{(E_6)_{\mathbb{C}} \times \mathbb{C}^{*}\}$

<u>or to</u>

(16.1.7) $\{\mathbb{C}^{10} + (\mathbb{C}^{2} \otimes \mathbb{C}^{16})\}\cdot\{SL(2;\mathbb{C}) \times Spin(10;\mathbb{C}) \times \mathbb{C}^{*}\}$

<u>or to</u>

(16.1.8) $\{\mathbb{C} + \mathbb{C}^{32}\}\cdot\{Spin(12;\mathbb{C}) \times \mathbb{C}^{*}\}$.

3. $G = E_{7,D_6A_1}$ <u>and</u> P <u>is isomorphic either to</u>

(16.1.9) $\{\mathbb{R}^{3,7} + (\mathbb{R}^{2} \otimes \mathbb{R}^{16})\}\cdot\{SU(2) \times Spin(3,7) \times \mathbb{R}^{+}\}$

<u>or to</u>

(16.1.10) $\{\mathbb{R} + \mathbb{R}^{32}\}\cdot\{Spin^{*}(12) \times \mathbb{R}^{+}\}$.

4. $G = E_{7,E_6T_1}$ <u>and</u> P <u>is isomorphic either to</u>

(16.1.11) $\mathbb{R}^{27}\cdot\{E_{6,F_4} \times \mathbb{R}^{+}\}$

<u>or to</u>

(16.1.12) $\{\mathbb{R}^{1,9} + (\mathbb{R}^{2} \otimes \mathbb{R}^{16})\}\cdot\{SL(2;\mathbb{R}) \times Spin(1,9) \times \mathbb{R}^{+}\}$

<u>or to</u>

(16.1.13) $\{\mathbb{R} + \mathbb{R}^{32}\}\cdot\{Spin(2,10) \times \mathbb{R}^{+}\}$.

16.2. Exceptional simple Jordan algebras. In order to analyse the cases (16.1.3), (16.1.6) and (16.1.11), in particular to understand the M-orbit structure on $\mathfrak{z}^* = \mathfrak{n}^*$ there, we need some facts about the structure and automorphisms groups of the exceptional simple Jordan algebras. References: McCrimmon [16] for a quick overview; Drucker [4] for real forms and for information that combines with [2] and [6] to give their structure groups; Jacobson ([6], [7]), Braun and Koecher [2], and Loos [15] for comprehensive discussions.

As before, $\mathfrak{C}_{\mathbb{C}}$ is the complex Cayley algebra. It has two real forms, the real Cayley division algebra $\mathfrak{C}$ and the split (has zero-divisors) real Cayley algebra $\mathfrak{C}_{\mathbb{R}}$. We write $x \mapsto \bar{x}$ for the conjugation of $\mathfrak{C}_{\mathbb{C}}$ over $\mathbb{C}$, or $\mathfrak{C}$ or $\mathfrak{C}_{\mathbb{R}}$ over $\mathbb{R}$.

The complex exceptional simple Jordan algebra $\mathfrak{J}_{\mathbb{C}}$ consists of all 3×3 "hermitian" matrices

$$(16.2.1) \qquad \underline{\underline{x}} = \begin{pmatrix} \xi_1 & x_3 & \bar{x}_2 \\ \bar{x}_3 & \xi_2 & x_1 \\ x_2 & \bar{x}_1 & \xi_3 \end{pmatrix}: \quad \xi_1, \xi_2, \xi_3 \in \mathbb{C} \quad \text{and} \quad x_1, x_2, x_3 \in \mathfrak{C}_{\mathbb{C}}$$

with product $\underline{\underline{x}} \cdot \underline{\underline{y}} = \frac{1}{2}\{\underline{\underline{xy}} + \underline{\underline{yx}}\}$. It has trace function $\mathrm{trace}(\underline{\underline{x}}) = \xi_1 + \xi_2 + \xi_3$ and norm function

$$(16.2.2) \quad \det(\underline{\underline{x}}) = \xi_1\xi_2\xi_3 - \xi_1 x_1\bar{x}_1 - \xi_2 x_2\bar{x}_2 - \xi_3 x_3\bar{x}_3 + (x_3x_1)x_2 + \bar{x}_2(\bar{x}_1\bar{x}_3).$$

The algebra $\mathfrak{J}_{\mathbb{C}}$ has three real forms,

$$(16.2.3) \quad \begin{cases} \mathfrak{J} : \text{all } \underline{\underline{x}} \in \mathfrak{J}_{\mathbb{C}} \text{ with } \xi_j \in \mathbb{R}, \quad x_j \in \mathfrak{C} \\ \mathfrak{J}_{\mathbb{R}} : \text{all } \underline{\underline{x}} \in \mathfrak{J}_{\mathbb{C}} \text{ with } \xi_j \in \mathbb{R}, \quad x_j \in \mathfrak{C}_{\mathbb{R}} \\ \mathfrak{J}_0 : \text{all } \underline{\underline{x}} \subset \mathfrak{J}_{\mathbb{C}} \text{ with } \xi_j \in \mathbb{R}; \quad x_2 \in \mathfrak{C};\ x_1, x_3 \in \sqrt{-1}\,\mathfrak{C} \end{cases}$$

called the "compact," the "split," and the "other" real exceptional simple Jordan algebra. The functions trace($\underline{\underline{x}}$) and det($\underline{\underline{x}}$) are real-valued on those real forms. See [4 ; Chapter I, §4].

The Cayley algebras have automorphism groups $\mathrm{Aut}(\mathfrak{C}_{\mathbb{C}}) = (G_2)_{\mathbb{C}}$, $\mathrm{Aut}(\mathfrak{C}) = G_2$ and $\mathrm{Aut}(\mathfrak{C}_{\mathbb{R}}) = G_{2,A_1A_1}$. Here the action on the pure imaginary $(\bar{x} = -x)$ Cayley numbers is 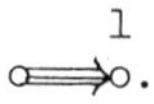.

The Jordan algebras have automorphism groups

$$(16.2.4)\qquad \begin{cases} \mathrm{Aut}(\mathfrak{J}_{\mathbb{C}}) = (F_4)_{\mathbb{C}}\,, & \mathrm{Aut}(\mathfrak{J}) = F_4, \\[2ex] \mathrm{Aut}(\mathfrak{J}_{\mathbb{R}}) = F_{4,C_1C_3}\,, & \mathrm{Aut}(\mathfrak{J}_0) = F_{4,B_4}\,. \end{cases}$$

See the $\mathfrak{A}_1^c$ column of the second chart on [4 , p. 38]. These fix the identity element $I = \begin{pmatrix} 1 & 0 & 0 \\ 0 & 1 & 0 \\ 0 & 0 & 1 \end{pmatrix}$, and act on a complement (trace($\underline{\underline{x}}$) = 0) to the span of I by the 26-dimensional representation 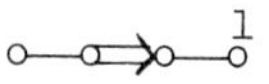 .

The structure group Str(•) of one of these Jordan algebras is defined to be the set of all invertible linear transformations T for which there exists a scalar $X(T)$ such that

$$(16.2.5)\qquad \det(T\underline{\underline{x}}) = X(T)\det(\underline{\underline{x}}), \quad \text{all } \underline{\underline{x}}.$$

Evidently X is a quasi-character on the structure group. The reduced structure group is the kernel of X. These reduced structure groups are given by

$$(16.2.6)\qquad \begin{cases} Str_1(\mathfrak{J}_{\mathbb{C}}) = (E_6)_{\mathbb{C}}\,, & Str_1(\mathfrak{J}) = E_{6,F_4}\,, \\ Str_1(\mathfrak{J}_{\mathbb{R}}) = E_{6,C_4}\,, & Str_1(\mathfrak{J}_0) = E_{6,F_4}\,. \end{cases}$$

See the $\mathfrak{A}_2^s$ column of the second chart on [4, p. 38] using [2; Satz 5.1 on p. 289] and the isomorphism of [4; p. 47]. They act irreducibly on the Jordan algebra by the 27-dimensional representation 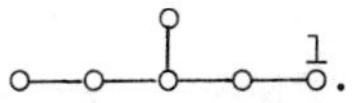. The isotropy subgroup at I is the automorphism group. The full structure group acts as the reduced structure group together with all nonzero scalar multiplications. The subgroup given by scalars of absolute value 1 is a sort of <u>semi-reduced structure group</u>

$$(16.2.7)\qquad M = \{T \in Str(\cdot) : |\chi(T)| = 1\}\,.$$

The semi-reduced structure group M of (16.2.7) acts on the real norm-hypersurface

$$(16.2.8)\qquad S = \{\underline{x} \text{ in the Jordan algebra: } |\det(\underline{x})| = 1\}\,.$$

We want the M-orbit structure of S. First, if $\underline{x} \in S$ then [2 ; Kap VII, Satz 7.2 on page 233] some $m \in M$ sends $\underline{x}$ to a diagonal matrix $\underline{y} = \begin{pmatrix} \eta_1 & 0 & 0 \\ 0 & \eta_2 & 0 \\ 0 & 0 & \eta_3 \end{pmatrix}$. Here $1 = |\det(\underline{y})| = |\eta_1\eta_2\eta_3|$. Second, as on [2, p. 233], M contains the transformations $\underline{z} \to \underline{wzw}$ where $\underline{w} = \begin{pmatrix} \omega_1 & 0 & 0 \\ 0 & \omega_2 & 0 \\ 0 & 0 & \omega_3 \end{pmatrix}$, ω_j scalars such that $|\omega_1\omega_2\omega_3| = 1$. This allows us to assume $\underline{y} = I$ in the complex case, $\underline{y} = \begin{pmatrix} \pm 1 & 0 & 0 \\ 0 & \pm 1 & 0 \\ 0 & 0 & \pm 1 \end{pmatrix}$ in the real cases. M also contains $\underline{z} \mapsto -\underline{z}$ and the $\underline{z} \mapsto \underline{wzw}$ where $\underline{w}$

is a permutation matrix of square I. Now we have

16.2.9. Lemma. In the complex case M is transitive on S. In the real cases, the only M-orbits are $M \cdot I$ and $M \cdot \underline{\underline{a}}$ where $\underline{\underline{a}} = \begin{pmatrix} -1 & 0 & 0 \\ 0 & -1 & 0 \\ 0 & 0 & 1 \end{pmatrix}$.

There is an equivalence relation on Jordan algebras called isotopy which induces isomorphisms of the structure groups. If $\mathcal{H}$ is the Jordan algebra and $\underline{\underline{a}} \in \mathcal{H}$ is invertible, then the isotope $(\mathcal{H}, \underline{\underline{a}})$ is the vector space $\mathcal{H}$ with multiplication twisted by $\underline{\underline{a}}$ in a certain way, and $\underline{\underline{a}}$ is the identity element of $(\mathcal{H}, \underline{\underline{a}})$. When $\mathcal{H}$ is $\mathcal{J}_{\mathbb{C}}$, $\mathcal{J}$ or $\mathcal{J}_{\mathbb{R}}$, 3×3 "hermitian" matrices over $\mathfrak{D} = \mathfrak{C}_{\mathbb{C}}$, $\mathfrak{C}$ or $\mathfrak{C}_{\mathbb{R}}$, the isotopes of interest to us are given as follows [6 ; Chapter I, Theorem 14 on p. 61]. Let $\underline{\underline{a}} \in \mathcal{H}$ be diagonal and invertible. Then $\underline{\underline{x}} \to \underline{\underline{xa}}$ is an isomorphism of $(\mathcal{H}, \underline{\underline{a}})$ onto

$$\mathcal{H}_3(\mathfrak{D}, j_{\underline{\underline{a}}}) = \{\underline{\underline{z}} \in \mathfrak{D}^{3\times 3} : (\underline{\underline{a}}^{-1})^t \bar{\underline{\underline{z}}}\underline{\underline{a}} = \underline{\underline{z}}\} \tag{16.2.10}$$

which carries the standard Jordan product.

16.2.11. Lemma. Let $\underline{\underline{a}} = \begin{pmatrix} -1 & 0 & 0 \\ 0 & -1 & 0 \\ 0 & 0 & 1 \end{pmatrix}$ as in Lemma 16.2.9. Then $(\mathcal{J}, \underline{\underline{a}}) \cong \mathcal{J}_0$, $(\mathcal{J}_0, \underline{\underline{a}}) \cong \mathcal{J}$ and $(\mathcal{J}_{\mathbb{R}}, \underline{\underline{a}}) \cong \mathcal{J}_{\mathbb{R}}$.

Proof. One computes $\mathcal{H}_3(\mathfrak{C}, j_{\underline{\underline{a}}}) \cong \mathcal{J}_0$; that shows $(\mathcal{J}, \underline{\underline{a}}) \cong \mathcal{J}_0$. As $\underline{\underline{a}} = \underline{\underline{a}}^{-1}$ it follows that $(\mathcal{J}_0, \underline{\underline{a}}) \cong ((\mathcal{J}, \underline{\underline{a}}), \underline{\underline{a}}^{-2}) \cong \mathcal{J}$.

The element $\underline{\underline{a}}$ is a square in $\mathcal{J}_{\mathbb{R}}$, say of $\begin{pmatrix} 0 & \alpha & 0 \\ \bar{\alpha} & 0 & 0 \\ 0 & 0 & 1 \end{pmatrix}$ where $\alpha \in \mathfrak{C}_{\mathbb{R}}$ with $\bar{\alpha} = -\alpha$ and $\alpha^2 = 1$. This implies [6, p. 60] that $(\mathcal{J}_{\mathbb{R}}, \underline{\underline{a}}) \cong \mathcal{J}_{\mathbb{R}}$. q.e.d.

16.2.12. Theorem Let $\underline{\underline{a}} = \begin{pmatrix} -1 & 0 & 0 \\ 0 & -1 & 0 \\ 0 & 0 & 1 \end{pmatrix}$ as above. Then the M-orbit structure of S is given as follows.

1. In $\mathcal{J}_{\mathbb{C}}$ and in $\mathcal{J}_{\mathbb{R}}$, M is transitive on S, with isotropy subgroup at I given by $\mathrm{Aut}(\mathcal{J}_{\mathbb{C}}) = (F_4)_{\mathbb{C}}$ in $\mathcal{J}_{\mathbb{C}}$, by $\mathrm{Aut}(\mathcal{J}_{\mathbb{R}}) = F_{4,C_1C_3}$ in $\mathcal{J}_{\mathbb{R}}$.

2. In $\mathcal{J}$ and in $\mathcal{J}_0$, there are two M-orbits on S, M(I) and $M(\underline{\underline{a}})$. The isotropy subgroups are

	at I	at $\underline{\underline{a}}$
in $\mathcal{J}$	$\mathrm{Aut}(\mathcal{J}) = F_4$	$\mathrm{Aut}(\mathcal{J}_0) = F_{4,B_4}$
in $\mathcal{J}_0$	$\mathrm{Aut}(\mathcal{J}_0) = F_{4,B_4}$	$\mathrm{Aut}(\mathcal{J}) = F_4$

Proof. The isotropy subgroup at I of the structure group is the automorphism group. In view of Lemma 16.2.9 we now need only prove

(i) in $\mathcal{J}_{\mathbb{R}}$, $\underline{\underline{a}} \in M(I)$;

(ii) in $\mathcal{J}$ and in $\mathcal{J}_0$, $\underline{\underline{a}} \notin M(I)$;

(iii) in $\mathcal{J}$ and in $\mathcal{J}_0$, the isotropy of M at $\underline{\underline{a}}$ is as claimed.

If $\mathcal{H}$ is one of $\mathcal{J}_{\mathbb{R}}$, $\mathcal{J}$ or $\mathcal{J}_0$, then $\underline{\underline{a}}$ is the identity element of $(\mathcal{H},\underline{\underline{a}})$, so the isotropy subgroup of M at $\underline{\underline{a}}$ is $\mathrm{Aut}(\mathcal{H},\underline{\underline{a}})$. Now (iii) follows from Lemma 16.2.11, and there $M_I \not\cong M_{\underline{\underline{a}}}$ implies (ii).

$\mathrm{Str}(\mathcal{J}_{\mathbb{R}})$ is just the group of self-isotopies of $\mathcal{J}_{\mathbb{R}}$, and Lemma 16.2.11 shows that it contains multiplication by $\underline{\underline{a}}$. Since $|\det \underline{\underline{a}}| = 1 = |\det I|$, the isotopy $\mathcal{J}_{\mathbb{R}} \to (\mathcal{J}_{\mathbb{R}},\underline{\underline{a}})$ belongs to M, so $\underline{\underline{a}} \in M(I)$ in $\mathcal{J}_{\mathbb{R}}$. q.e.d.

16.3. $G = E_{7,A_7}$ and P is given by (16.1.3).

Here $Z = \mathbb{R}^{27} = \mathfrak{z}$. Using (16.2.6) we identify $\mathfrak{z}^*$ with the split exceptional simple real Jordan algebra $\mathcal{J}_{\mathbb{R}}$; this carries the action of MA over to the action of the structure group $\mathrm{Str}(\mathcal{J}_{\mathbb{R}})$, sending M to the semi-reduced structure group (16.2.7). Also, identify $\mathfrak{z}$ and Z with $\mathcal{J}_{\mathbb{R}}$ under $\lambda(z) = \mathrm{trace}\,(\lambda z)$.

The M-invariant polynomial is

$$(16.3.1) \qquad \psi(\lambda) = |\det(\lambda)|^2, \qquad \det \text{ taken in } \mathcal{J}_{\mathbb{R}} \text{ via (16.2.2).}$$

The data k, ℓ, q of (9.3a) are 27, 0, 27, and ψ has degree 6, so our positive self-adjoint invertible operator D on $L^2(P)$ is

$$(16.3.2) \qquad D = \psi^{9/2}, \qquad \text{not differential.}$$

From (16.2.6) and Theorem 16.2.12, $M \cong E_{6,C_4} \times \mathbb{Z}_4$, and M is transitive on the "unit sphere" $S = \{\lambda \in \mathfrak{z}^* : \psi(\lambda) = 1\}$ with isotropy subgroup F_{4,C_1C_3} at I. Let $[\pi] \in \hat{N}$ denote the unitary representation class specified by I,

$$\pi(z) = e^{\sqrt{-1}\,\mathrm{trace}(z)}.$$

Since $F_{4,C_1C_3} = \mathrm{Aut}(\mathcal{J}_{\mathbb{R}})$ preserves the trace function, $[\pi]$ extends to a class $[\tilde{\pi}] \in (N \cdot F_{4,C_1C_3})^{\wedge}$. Now the "generic" representations of P are the

$$(16.3.3) \qquad \pi_\nu = \mathrm{Ind}_{N \cdot F_{4,C_1C_3}\uparrow P}(\tilde{\pi} \otimes \nu), \quad [\nu] \in (F_{4,C_1C_3})^{\wedge}.$$

That leads to the Fourier Inversion formula

$$(16.3.4) \qquad F(1_P) = \int_{(F_{4,C_1C_3})^\wedge} \text{trace}\ \pi_\nu(DF)d\mu(\nu)$$

for Schwartz class functions F on P that satisfy (9.5).

<u>16.4</u>. $G = E_{7,A_7}$ <u>and</u> P <u>is given by</u> (16.1.4).

Here $Z = \mathbb{R}^{5,5} = \mathfrak{z}$, we identify $\mathfrak{z}^*$ with $\mathbb{R}^{5,5}$ under $\lambda(z) = \langle\lambda,z\rangle$, and $M = SL(2;\mathbb{R}) \times Spin(5,5)$ acts on $\mathfrak{z}^*$ by $(\gamma,g): \lambda \mapsto \nu(g)\lambda$ where ν: 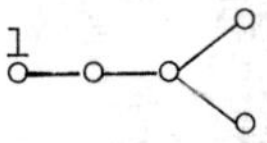 is the vector representation. The M-invariant polynomial is

$$(16.4.1) \qquad \psi(\lambda) = \langle\lambda,\lambda\rangle \quad \text{on} \quad \mathfrak{z}^* \cong \mathbb{R}^{5,5}, \quad \text{degree} \quad d = 2.$$

The data k,ℓ,q of (9.3a) are 10, 32, 26; so we have a choice of self-adjoint operators D on $L^2(P)$,

$$(16.4.2a) \qquad D = \square^{13}, \quad \text{differential but not positive}$$

$$(16.4.2b) \qquad D = |\square|^{13}, \quad \text{positive but not differential}$$

where $\square$ is the Laplacian on $\mathbb{R}^{5,5}$.

The "unit sphere" $S = \{\lambda \in \mathfrak{z}^* : \langle\lambda,\lambda\rangle = \pm 1\}$ breaks into two orbits

$$(16.4.3) \qquad S_\pm = Ad^*(M)\cdot\lambda_\pm \cong M/\{SL(2;\mathbb{R}) \times Spin(4,5)\}$$

where $\langle \lambda_{\pm}, \lambda_{\pm} \rangle = \pm 1$. Let $[\pi_{\pm}] \in \hat{N}$ denote the unitary representation classes specified by $\lambda_{\pm}$.

16.4.4. Lemma. The isotropy subgroups $M_{\pm} \cong SL(2;\mathbb{R}) \times Spin(4,5)$ of M at $\lambda_{\pm}$ satisfy $H^2(M_{\pm};\mathbb{C}') \cong \mathbb{C}' \times \mathbb{Z}_2$. Let σ denote the element of order 2 in the circle factor $\mathbb{C}'$. Then σ is the Mackey obstruction to extending $\pi_{\pm}$ from N to $NM_{\pm}$, that is $[\pi_{\pm}]$ extend to σ-representation classes $[\tilde{\pi}_{\pm}] \in (NM_{\pm})^{\wedge}_{\sigma}$.

Proof. For the assertion on $H^2(M_{\pm};\mathbb{C}')$ note $\pi_1(SL(2;\mathbb{R})) \cong \mathbb{Z}$ and $\pi_1(Spin(4,5)) \cong \mathbb{Z}_2$. If $f: M_{\pm} \to Sp(16;\mathbb{R})$ denotes the representation on N/Z, now $f_{\#}: \pi_1(M_{\pm}) \to \pi_1(Sp(16;\mathbb{R}))$ kills the finite group $\mathbb{Z}_2$, and [28 , Lemma 9.18(2)] $f_{\#}$ maps $\pi_1(SL(2;\mathbb{R}))$ isomorphically to $\pi_1(Sp(16;\mathbb{R}))$. The assertion now follows that σ, f^*-image of the element of order 2 in $H^*(Sp(16;\mathbb{R}); \mathbb{C}')$, is the Mackey obstruction in question.

q.e.d.

Now the "generic" representations of P are the

$$(16.4.5) \qquad \pi_{\pm,\nu} = \mathrm{Ind}_{NM_{\pm}\uparrow P}(\tilde{\pi}_{\pm} \otimes \nu), \; [\nu] \in (M_{\pm})^{\wedge}_{\sigma} .$$

That leads to the Fourier Inversion formula

$$(16.4.6) \quad F(1_P) = \int_{\{SL(2;\mathbb{R})\times Spin(4,5)\}^{\wedge}_{\sigma}} \{\mathrm{trace}\ \pi_{+,\nu}(DF)+\mathrm{trace}\ \pi_{-,\nu}(DF)\}d\mu(\nu)$$

where $F: P \to \mathbb{C}$ is of Schwartz class, and further restricted in partial Fourier transform in case we make the choice (16.4.2b) for D.

16.5. $G = E_{7,A_7}$ and P is given by (16.1.5).

Here $Z = \mathbb{R} = \mathfrak{z}$, and $\mathfrak{z}^* \cong \mathbb{R}$ under $\lambda(z) = \lambda z$. $M = Spin(6,6)$ acts trivially on $\mathfrak{z}^*$ so we use the invariant polynomial

(16.5.1) $\psi(\lambda) = \lambda$, degree $d = 1$.

The data k, ℓ, q of (9.3a) are 1, 32,17; so we have a choice of self-adjoint operator D on $L^2(P)$,

(16.5.2a) $D = (\partial/\partial z)^{17}$, differential but not positive,

(16.5.2b) $D = |\partial/\partial z|^{17}$, positive but not differential.

The "unit sphere" $S = \{\pm 1\} \subset \mathfrak{z}^*$ consists of two points fixed under M. Let $[\pi_{\pm}]$ denote the corresponding classes in $\hat{N}$. Since $\pi_1(M) \cong \mathbb{Z}_2$, finite, they extend to unitary representation classes $[\widetilde{\pi_{\pm}}] \in (NM)^{\wedge}$. The "generic" representations of P now are the

(16.5.3) $\pi_{\pm,\nu} = \mathrm{Ind}_{NM\uparrow P}(\widetilde{\pi_{\pm}} \otimes \nu)$, $[\nu] \in \hat{M}$.

Thus we come to the Fourier Inversion formula

(16.5.4) $$F(1_P) = \int_{Spin(6,6)^{\wedge}} \{\text{trace } \pi_{+,\nu}(DF) + \text{trace } \pi_{-,\nu}(DF)\} d\mu(\nu)$$

for Schwartz class functions $F\colon P \to \mathbb{C}$. Here [9 , Theorem 2.7], applied to $(-\partial^2/\partial z^2)^{17/2}$, shows that there is no further restriction on F in case we choose $D = |\partial/\partial z|^{17}$.

<u>16.6</u>. $G = (E_7)_{\mathbb{C}}$ <u>and</u> P <u>is given by</u> (16.1.6).

Here $Z = \mathbb{C}^{27} = \mathfrak{z}$. We use (16.2.6) to identify $\mathfrak{z}^*$ with the complex exceptional simple Jordan algebra $\mathcal{J}_{\mathbb{C}}$; this carries the action of MA to $\mathrm{Str}(\mathcal{J}_{\mathbb{C}})$, the action of $M \cong (E_6)_{\mathbb{C}} \cdot U(1)$ to the semi-reduced structure group (16.2.7). Also, identify Z and $\mathfrak{z}$ with $\mathcal{J}_{\mathbb{C}}$ under $\lambda(\mathbf{z}) = \mathrm{Re\ Trace}(\lambda\mathbf{z})$.

The M-invariant polynomial is

$$\psi(\lambda) = |\det(\lambda)|^2, \quad \det \text{ for } \mathcal{J}_{\mathbb{C}} \text{ as in (16.2.2).} \tag{16.6.1}$$

The data k, ℓ, q of (9.3a) are 54, 0, 54, so our invertible positive self-adjoint operator D on $L^2(P)$ is

$$D = \Psi^9, \quad \text{differential.} \tag{16.6.2}$$

Theorem 16.2.12 shows that M is transitive on the "unit sphere" $S = \{\lambda \in \mathfrak{z}^* : \psi(\lambda) = 1\}$ with isotropy subgroup $(F_4)_{\mathbb{C}}$ at I. Let $[\pi] \in \hat{N}$ correspond to $I \in S$,

$$\pi(z) = e^{\sqrt{-1}\,\mathrm{Re\ Trace}(\dot{z})} .$$

As $(F_4)_{\mathbb{C}} = \mathrm{Aut}(\mathcal{J}_{\mathbb{C}})$ preserves the trace function, and in any case is a simply connected semisimple Lie group, $[\pi]$ extends to a class $[\tilde{\pi}] \in \{N \cdot (F_4)_{\mathbb{C}}\}^{\wedge}$. So the "generic" representations of P are the

$$\pi_\nu = \mathrm{Ind}_{N\cdot(F_4)_{\mathbb{C}} \uparrow P} (\tilde{\pi} \otimes \nu), \quad [\nu] \in \{(F_4)_{\mathbb{C}}\}^{\wedge} . \tag{16.6.3}$$

That leads to the Fourier Inversion formula

$$F(1_P) = \int_{\{(F_4)_{\mathbb{C}}\}^\wedge} \operatorname{trace} \pi_\nu(DF)\,d\mu(\nu) \tag{16.6.4}$$

for Schwartz class functions $F\colon P \to \mathbb{C}$.

<u>16.7</u>. $G = (E_7)_{\mathbb{C}}$ <u>and</u> P <u>is given by</u> (16.1.7).

Here $Z = \mathbb{C}^{10} = \mathfrak{z}$, and $\mathfrak{z}^* \cong \mathbb{C}^{10}$ under $\lambda(z) = \mathrm{Re}\langle \lambda, z\rangle$ where $\langle\ ,\ \rangle$ is the $SO(10;\mathbb{C})$-invariant bilinear form. Thus $M = SL(2;\mathbb{C}) \times Spin(10;\mathbb{C}) \times \mathbb{C}'$ acts on $\mathfrak{z}^*$ by $(\gamma, g, t)\colon \lambda \mapsto t^{-2}\cdot\nu(g)\lambda$ where ν: 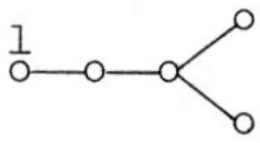 is the vector representation. Our M-invariant $\mathbb{R}$-polynomial on $\mathfrak{z}^*$ is

$$\psi(\lambda) = |\langle \lambda, \lambda\rangle|^2, \quad \text{degree } d = 4\ . \tag{16.7.1}$$

The data k, ℓ, q of (9.3a) are 20, 64, 52; so our positive self-adjoint operator on $L^2(P)$ is

$$D = \Psi^{13}, \quad \text{differential.} \tag{16.7.2}$$

Note that Ψ corresponds to the square of the Laplacian of $\mathbb{R}^{10,10}$ under $(\mathbb{C}^{10}, \langle\ ,\ \rangle) \leftrightarrow \mathbb{R}^{10,10}$.

The "unit sphere" $S = \{\lambda \in \mathfrak{z}^*\colon |\langle \lambda, \lambda\rangle| = 1\}$ is a single M-orbit,

$$S = Ad^*(M)\cdot\lambda_1 \cong M/\{SL(2;\mathbb{C}) \times Spin(9;\mathbb{C}) \times \{\pm 1\}\}\ . \tag{16.7.3}$$

Let $[\pi] \in \hat{N}$ correspond to $\lambda_1 \in \mathfrak{z}^*$. The isotropy subgroup M_1 of M at λ_1 has $\pi_1(M_1)$ finite, so $[\pi]$ extends to a class $[\tilde{\pi}] \in (NM_1)\hat{}$. Now the "generic" representations of P are the

$$(16.7.4) \qquad \pi_\nu = \mathrm{Ind}_{NM_1 \uparrow P}(\tilde{\pi} \otimes \nu), \qquad [\nu] \in \widehat{M_1}.$$

That leads to the Fourier Inversion formula

$$(16.7.5) \qquad F(1_P) = \int_{\{SL(2;\mathbb{C})\times Spin(9;\mathbb{C})\times \mathbf{Z}_2\}\hat{}} \text{trace } \pi_\nu(DF)d\mu(\nu)$$

for Schwartz class functions $F: P \to \mathbb{C}$.

16.8. $G = (E_7)_{\mathbb{C}}$ and P is given by (16.1.8).

Here $Z = \mathbb{C} = \mathfrak{z}$, and $\mathfrak{z}^* \cong \mathbb{C}$ under $\lambda(z) = \mathrm{Re}(\lambda z)$. $M = Spin(12;\mathbb{C}) \times \mathbb{C}'$ acts on $\mathfrak{z}^*$ by $(g,t): \lambda \mapsto t^{-2}\lambda$. Our M-invariant $\mathbb{R}$-polynomial on $\mathfrak{z}^*$ is

$$(16.8.1) \qquad \psi(\lambda) = |\lambda|^2, \qquad \text{degree } d = 2.$$

The data of (9.3a) are 2, 64, 34; so our positive self-adjoint operator D on $L^2(P)$ is

$$(16.8.2) \qquad D = \Delta^{17}, \qquad \Delta \text{ real Laplacian on } \mathbb{C} \cong \mathbb{R}^2.$$

The "unit sphere" $S = \{\lambda \in \mathfrak{z}^*: |\lambda| = 1\}$ is a circle on which the $\mathbb{C}'$-factor of M is transitive,

(16.8.3) $$S = \mathrm{Ad}^*(M)\cdot\lambda_1 \cong M/\{\mathrm{Spin}(12;\mathbb{C}) \times \mathbb{Z}_2\}.$$

The class $[\pi] \in \hat{N}$, corresponding to λ_1, extends to a class $[\tilde{\pi}] \in (NM_1)^\wedge$ because $\pi_1(M_1)$ is finite. So we have "generic" representations of P given by

(16.8.4) $$\pi_\nu = \mathrm{Ind}_{NM_1\uparrow P}(\tilde{\pi} \otimes \nu), \quad [\nu] \in \widehat{M_1}.$$

That leads directly to the Fourier Inversion formula

(16.8.5) $$F(1_P) = \int_{\{\mathrm{Spin}(12;\mathbb{C}) \times \mathbb{Z}_2\}^\wedge} \mathrm{trace}\ \pi_\nu(DF)d\mu(\nu)$$

for Schwartz class functions $F\colon P \to \mathbb{C}$.

<u>16.9</u>. $G = E_{7,D_6A_1}$ <u>and</u> P <u>is given by</u> (16.1.9).

Here $Z = \mathbb{R}^{3,7} = \mathfrak{z}$, and $\mathfrak{z}^* \cong \mathbb{R}^{3,7}$ under $\lambda(z) = \langle \lambda, z \rangle$. $M = SU(2) \times \mathrm{Spin}(3,7)$ acts on $\mathfrak{z}^*$ by $(\gamma,g)\colon \lambda \mapsto \nu(g)\lambda$ where $\nu\colon \mathrm{Spin}(3,7) \to SO(3,7)$ is the vector representation. Our M-invariant polynomial on $\mathfrak{z}^*$ is

(16.9.1) $$\psi(\lambda) = \langle \lambda, \lambda \rangle, \quad \text{degree } d = 2.$$

The data of (9.3a) are 10, 32, 26; so our invertible self-adjoint operator D on $L^2(P)$ can be taken to be either of

(16.9.2a) $$D = \square^{13}, \quad \text{differential but not positive}$$

(16.9.2b) $$D = |\square|^{13}, \quad \text{positive but not differential}$$

where $\square$ is the Laplacian of $\mathbb{R}^{3,7}$.

The "unit sphere" $S = \{\lambda \in \mathfrak{z}^*: \langle \lambda,\lambda \rangle = \pm 1\}$ consists of two orbits

$$(16.9.3) \quad \begin{cases} Ad^*(M)\cdot\lambda_+ \cong M/\{SU(2) \times Spin(2,7)\} \\ \\ Ad^*(M)\cdot\lambda_- \cong M/\{SU(2) \times Spin(3,6)\} \end{cases}$$

where $\langle \lambda_\pm,\lambda_\pm\rangle = \pm 1$. Let $[\pi_\pm] \in \hat{N}$ be the class specified by $\lambda_\pm$, and write $M_\pm$ for the isotropy subgroup of M at $\lambda_\pm$.

<u>16.9.4. Lemma</u>. <u>The classes</u> $[\pi_\pm] \in \hat{N}$ <u>extend to classes</u> $[\widetilde{\pi_\pm}] \in (NM_\pm)^\wedge$.

<u>Proof</u>. Choose a sign $\pm$, and let M' denote Spin(2,7) or Spin(3,6) so that $M_\pm = SU(2) \times M'$. The representation of $M_\pm$ on $N/Z = \mathbb{R}^2 \otimes \mathbb{R}^{16}$ splits as $f_1 \otimes f_2$. Let $T = \left\{t_\theta = \begin{pmatrix} e^{i\theta} & 0 \\ 0 & e^{-i\theta}\end{pmatrix}\right\} \subset SU(2)$. Then $(f_1 \otimes f_2)(t_\theta,1)$ has $e^{\pm i\theta}$ for eigenvalues, each with multiplicity 16. As in [21, p. 22], now the centralizer of $(f_1 \otimes f_2)(T \times \{1\})$ in $Sp(16;\mathbb{R})$ is a unitary group $U(p,q)$, $p + q = 16$. The case $\mathbb{F} = \mathbb{C}$ of [27 , Proposition 4.16] can be interpreted as saying that the element $\tilde{\sigma}$ of order 2 in $H^2(Sp(16;\mathbb{R});\mathbb{C}')$ restricts to zero in $H^2(U(p,q);\mathbb{C}')$, in particular restricts to zero in $H^2((f_1 \otimes f_2)(\{1\} \times M');\mathbb{C}')$. As $SU(2)$ is simply connected, now

$$\sigma = (f_1 \otimes f_2)^*(\tilde{\sigma}) \in H^2(M_\pm;\mathbb{C}')$$

vanishes. But σ is the Mackey obstruction to extending $[\pi_\pm]$. <u>q.e.d.</u>

Now the "generic" representations of P are the

$$(16.9.5) \qquad \pi_{\pm,\nu} = \mathrm{Ind}_{NM_{\pm}\uparrow P}(\widetilde{\pi_{\pm}} \otimes \nu), \quad [\nu] \in \widehat{M_{\pm}}.$$

Thus we are led to the Fourier Inversion formula

$$(16.9.6) \qquad \begin{cases} F(1_P) = \displaystyle\int_{\{SU(2)\times Spin(2,7)\}^{\wedge}} \mathrm{trace}\ \pi_{+,\nu}(DF)d\mu_{+}(\nu) \\ \qquad\qquad + \displaystyle\int_{\{SU(2)\times Spin(3,6)\}^{\wedge}} \mathrm{trace}\ \pi_{-,\nu}(DF)d\mu_{-}(\nu) \end{cases}$$

for Schwartz class functions $F: P \to \mathbb{C}$ which, in case of the choice (16.9.2b), also satisfy the additional condition on partial Fourier transform specified in (9.5).

16.10. $G = E_{7,D_6A_1}$ and P is given by (16.1.10).

Here $Z = \mathbb{R} = \mathfrak{z}$ and $\mathfrak{z}^* \cong \mathbb{R}$ under $\lambda(z) = \lambda z$. $M = Spin^*(12)$ acts trivially on $\mathfrak{z}^*$ so we use the invariant polynomial

$$(16.10.1) \qquad \psi(\lambda) = \lambda, \qquad \text{degree } d = 1,$$

and come to the choice of invertible self-adjoint operator on $L^2(P)$,

$$(16.10.2a) \qquad D = (\partial/\partial z)^{17}, \qquad \text{differential but not positive},$$

$$(16.10.2b) \qquad D = |\partial/\partial z|^{17}, \qquad \text{positive but not differential}.$$

The "unit sphere" $S = \{\pm 1\} \subset \mathfrak{z}^*$, two M-fixed points. Let $[\pi_\pm] \in \hat{N}$ denote the class specified by $\pm 1 \in S$.

16.10.3. Lemma. $H^2(M;\mathbb{C}')$ is a circle group. Let σ denote its element of order 2. Then $[\pi_\pm]$ extends to a cocycle representation class $[\widetilde{\pi_\pm}] \in (NM)^\wedge_\sigma$.

Proof. Here $M = \mathrm{Spin}^*(12)$ is the 2-sheeted covering group of $SO^*(12)$, the real form of $SO(12;\mathbb{C})$ with maximal compact subgroup $U(6)$. The action $f: M \to Sp(16;\mathbb{R})$ of M on N/Z is the half spin representation, highest weight $\frac{1}{2}(\varepsilon_1+\ldots+\varepsilon_6)$ in the diagram, where $\{\psi_1,\ldots,\psi_5\}$ are the simple roots for the maximal compact subgroup $\tilde{U}(6)$ and $\{\varepsilon_1,\ldots,\varepsilon_6\}$ are the weights of the vector representation $\tilde{U}(6) \to U(6) \subset GL(6;\mathbb{C})$. Now view f as a map of maximal compact subgroups,

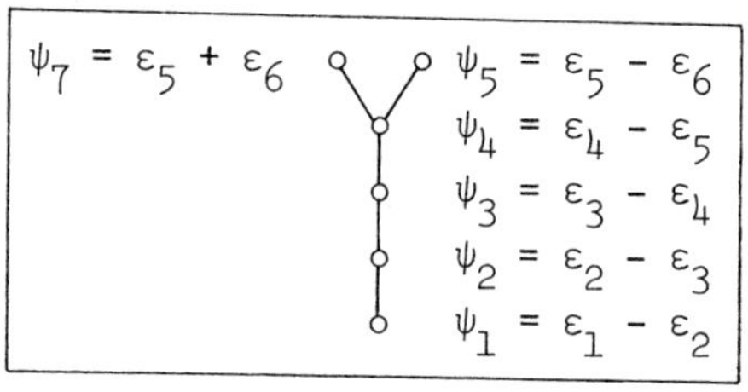

$$f: \tilde{U}(6) \to U(16).$$

As representation, it is direct sum of the unitary character $\exp(\frac{1}{2}(\varepsilon_1+\ldots+\varepsilon_6))$ and the 15-dimensional representation of highest weight

$$\frac{1}{2}(\varepsilon_1+\ldots+\varepsilon_6) - (\varepsilon_5+\varepsilon_6) = \frac{1}{2}(\varepsilon_1+\varepsilon_2+\varepsilon_3+\varepsilon_4-\varepsilon_5-\varepsilon_6)$$

which is 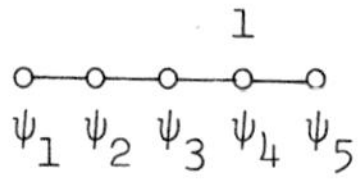

on the semisimple part of $\tilde{U}(6)$.

Let $\{r(\theta)\}$ denote the (central) 1-parameter subgroup of $\tilde{U}(6)$ such that $r(\theta)$ lies over $e^{i\theta}I_6 \in U(6)$. We have just seen that $f: \tilde{U}(6) \to U(16)$ satisfies

$$f(r(\theta)) = \begin{pmatrix} e^{3i\theta} & 0 \\ 0 & e^{i\theta}I_{15} \end{pmatrix} = e^{9i\theta/8} \begin{pmatrix} e^{15i\theta/8} & 0 \\ 0 & e^{-i\theta/8}I_{15} \end{pmatrix}.$$

Now, in Proposition 13.2.6, we have $n = 16$, $q = 9$, $h = 6$, and m such that $m\theta/8$ is parameter of period 2π in $\{r(\theta)\}$.

Since Spin*(12) has center $\mathbb{Z}_2 \times \mathbb{Z}_2$, and SO*(12) has center $\{\pm I_{12}\}$ and fundamental group $\mathbb{Z}$, the fundamental groups of groups covered by Spin*(12) are

$$\begin{array}{ll} \pi_1(\mathrm{Spin}^*(12)) \cong \mathbb{Z}, & \text{say with generator } a^2 \\ \pi_1(SO^*(12)) \cong \mathbb{Z}, & \text{with generator } a \\ \pi_1(SO^*(12)/\{\pm I\}) \cong \mathbb{Z} \times \mathbb{Z}_2, & \text{generators } a \text{ and } b \\ \pi_1(M/\ker(f)) \cong \mathbb{Z} \times \mathbb{Z}_2, & \text{generators } a^2 \text{ and } b . \end{array}$$

So $\{r(\theta)\}$ has the same period as $\{f(r(\theta))\}$ and $m = 8$ in Proposition 13.2.6. Now $r = nq/mh = 3$, and our assertion follows from Proposition 13.2.2. <u>q.e.d.</u>

Now the "generic" representations of P are the

$$(16.10.4) \qquad \pi_{\pm,\nu} = \mathrm{Ind}_{NM \dagger P}(\widetilde{\pi_\pm} \otimes \nu), \qquad [\nu] \in \hat{M}_\sigma .$$

This leads to the Fourier Inversion formula

(16.10.5) $$F(1_P) = \int_{\widehat{\mathrm{Spin}^*(12)}_\sigma} \{\mathrm{trace}\ \pi_{+,\nu}(DF) + \mathrm{trace}\ \pi_{-,\nu}(DF)\} d\mu(\nu)$$

where $F: P \to \mathbb{C}$ is a function of Schwartz class, [9 , Theorem 2.7] showing that the choice (16.10.2) imposes no further restriction on F.

16.11. $G = E_{7,E_6T_1}$ and P is given by (16.1.11).

Here $Z = \mathbb{R}^{27} = \mathfrak{z}$. Following (16.2.6) we identify $\mathfrak{z}^*$ with compact (= formally real) exceptional simple real Jordan algebra $\mathcal{J}$; this carries MA to $\mathrm{Str}(\mathcal{J})$ and carries $M \cong E_{6,F_4} \times \mathbb{Z}_2$ to the semi-reduced structure group (16.2.7). Identify Z and $\mathfrak{z}$ with $\mathcal{J}$ under $\lambda(z) = \mathrm{trace}(\lambda z)$.

The M-invariant polynomial is

(16.11.1) $$\psi(\lambda) = |\det(\lambda)|^2 , \quad \det \text{ for } \mathcal{J} \text{ as in (16.2.2).}$$

As in §16.3, now our invertible positive self-adjoint operator D on $L^2(P)$ is

(16.11.2) $$D = \Psi^{9/2} , \qquad \text{not differential.}$$

Theorem 16.2.12 shows that the "unit sphere" $S = \{\lambda \in \mathfrak{z}^* : \psi(\lambda) = 1\}$ is the union of two M-orbits,

$$(16.11.3)\quad \begin{cases} S_+ = Ad^*(M)\cdot I \cong (E_{6,F_4} \times \mathbb{Z}_2)/F_4 , \\ S_- = Ad^*(M)\cdot \underline{\underline{a}} \cong (E_{6,F_4} \times \mathbb{Z}_2)/F_{4,B_4} . \end{cases}$$

Let $[\pi_+]$, $[\pi_-] \in \hat{N}$ correspond to $I, \underline{\underline{a}} \in S$,

$$\pi_+(z) = e^{\sqrt{-1}\,\mathrm{trace}(z)} \text{ and } \pi_-(z) = e^{\sqrt{-1}\,\mathrm{trace}(\underline{\underline{a}}z)} .$$

The isotropy groups F_4 and F_{4,B_4} are simply connected semisimple Lie groups, so we have extensions

$$[\tilde{\pi}_+] \in (N\cdot F_4)^\wedge \text{ and } [\tilde{\pi}_-] \in (N\cdot F_{4,B_4})^\wedge .$$

Now the "generic" representations of P are the

$$(16.11.4)\qquad \pi_{+,\nu} = \mathrm{Ind}_{N\cdot F_4 \uparrow P}(\tilde{\pi}_+ \otimes \nu) \quad , \quad [\nu] \in \widehat{F_4}$$

and the

$$(16.11.5)\qquad \pi_{-,\nu} = \mathrm{Ind}_{N\cdot F_{4,B_4} \uparrow P}(\tilde{\pi}_- \otimes \nu), \quad [\nu] \in (F_{4,B_4})^\wedge .$$

That leads to the Fourier Inversion formula

$$(16.11.6)\quad \begin{cases} F(1_P) = \sum_{[\nu]\in(F_4)^\wedge} \mathrm{trace}\, \pi_{+,\nu}(DF)\dim(\nu) \\ \qquad + \int_{(F_{4,B_4})^\wedge} \mathrm{trace}\, \pi_{-,\nu}(DF)d\mu(\nu) \end{cases}$$

for Schwartz class functions $F\colon P \to \mathbb{C}$ that satisfy (9.5).

<u>16.12</u>. $G = E_{7,E_6T_1}$ <u>and</u> P <u>is given by</u> (16.1.12).

Here $Z = \mathbb{R}^{1,9} = \mathfrak{z}$, and $\mathfrak{z}^* \cong \mathbb{R}^{1,9}$ under $\lambda(z) = \langle \lambda, z \rangle$. $M = SL(2;\mathbb{R}) \times Spin(1,9)$ acts on $\mathfrak{z}^*$ by $(\gamma, g)\colon \lambda \mapsto \nu(g)\lambda$ where ν : 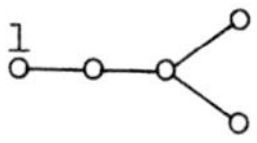 . The invariant polynomial is

$$\psi(\lambda) = \langle \lambda, \lambda \rangle \quad , \qquad \text{degree } d = 2. \tag{16.12.1}$$

That leads to a choice of self-adjoint operators,

(16.12.2a) $D = \square^{13}$, differential but not positive

(16.12.2b) $D = |\square|^{13}$, positive but not differntial

where $\square$ is the Laplacian on $\mathbb{R}^{1,9}$.

The "unit sphere" $S = \{\lambda \in \mathfrak{z}^*\colon \langle \lambda, \lambda \rangle = \pm 1\}$ falls into two M-orbits,

$$\begin{cases} Ad^*(M)\cdot\lambda_+ \cong M/\{SU(2) \times Spin(9)\} , \\ Ad^*(M)\cdot\lambda_- \cong M/\{SU(2) \times Spin(1,8)\} . \end{cases} \tag{16.12.3}$$

Let $[\pi_\pm] \in \hat{N}$ be the class specified by $\lambda_\pm$, write $M_\pm$ for the isotropy subgroup of M at $\lambda_\pm$. The argument of Lemma 16.9.4 -- or just finiteness of $\pi_1(M_\pm)$ -- shows that $[\pi_\pm]$ extends to a class $[\tilde{\pi}_\pm] \in (NM_\pm)^\wedge$. Now P has "generic" representations

(16.12.4) $\pi_{\pm,\nu} = \mathrm{Ind}_{NM_{\pm}\uparrow P}(\widetilde{\pi_{\pm}} \otimes \nu), \qquad [\nu] \in \widehat{M_{\pm}}$,

and we are led to the Fourier Inversion formula

(16.12.5) $$\begin{cases} F(1_P) = \sum_{\{SU(2)\times Spin(9)\}^\wedge} \operatorname{trace} \pi_{+,\nu}(DF)\dim(\nu) \\ \qquad + \int_{\{SU(2)\times Spin(1,8)\}^\wedge} \operatorname{trace} \pi_{-,\nu}(DF)d\mu_-(\nu) \end{cases}$$

for Schwartz functions F on P, further restricted in case of the choice (16.12.2b).

<u>16.13</u>. $G = E_{7,E_6T_1}$ <u>with</u> P <u>given by</u> (16.1.13).

Here $Z = \mathbb{R} = \mathfrak{z}$, so $\mathfrak{z}^* \cong \mathbb{R}$ under $\lambda(z) = \lambda z$, and $M = Spin(2,10)$ acts trivially on $\mathfrak{z}^*$. So we come to

(16.13.1) $\psi(\lambda) = \lambda$, degree $d = 1$

and our choice of

(16.13.2a) $D = (\partial/\partial z)^{17}$, differential but not positive

(16.13.2b) $D = |\partial/\partial z|^{17}$, positive but not differential.

The "unit sphere" $S = \{\pm 1\} \subset \mathfrak{z}^*$, two M-fixed points. Let $[\pi_{\pm}] \in \hat{N}$ denote the corresponding representation classes.

16.13.3. Lemma. Each $[\pi_\pm] \in \hat{N}$ extends to a class $[\widetilde{\pi_\pm}] \in (NM)\hat{}$.

Proof. $M = \mathrm{Spin}(2,10)$ acts on N/Z by a half spin representation $f\colon M \to Sp(16;\mathbb{R})$. Since $\mathrm{Spin}(2,10)$ has center $\mathbb{Z}_2 \times \mathbb{Z}_2$, $SO(2,10)$ has center $\mathbb{Z}_2$, and $SO(2,10)$ has fundamental group $\mathbb{Z} \times \mathbb{Z}_2$,

$$\begin{array}{ll}
\pi_1(\mathrm{Spin}(2,10)) \cong \mathbb{Z}, & \text{say with generator } a^2 \\
\pi_1(SO(2,10)) \cong \mathbb{Z} \times \mathbb{Z}_2, & \text{with generators } a^2, b \\
\pi_1(SO(2,10)/\{\pm I\}) \cong \mathbb{Z} \times \mathbb{Z}_2, & \text{with generators } a, b \\
\pi_1(M/\ker(f)) \cong \mathbb{Z}, & \text{with generator } a .
\end{array}$$

Now we have

$$\begin{array}{ccc}
M = \mathrm{Spin}(2,10) & & \\
\downarrow p & \searrow^{f} & \\
M/\ker(f) & \xrightarrow{\ \bar{f}\ } & Sp(16;\mathbb{R})
\end{array}$$

in such a way that: if $\bar{f}_\# \pi_1(M/\ker(f))$ has index r' in $\pi_1(Sp(16;\mathbb{R}))$, then $f_\# \pi_1(M)$ has index $r = 2r'$. The assertion now follows from Proposition 13.2.2.

q.e.d.

Now the "generic" representations of P are the

$$(16.13.4) \qquad \pi_{\pm,\nu} = \mathrm{Ind}_{NM\uparrow P}(\widetilde{\pi_\pm} \otimes \nu), \qquad [\nu] \in \hat{M}.$$

This leads to the Fourier Inversion formula

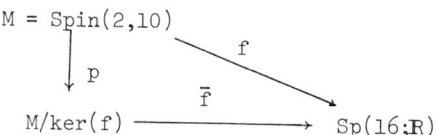

$$(16.13.5)\quad F(1_P) = \int_{\mathrm{Spin}(2,10)^\wedge} \{\mathrm{trace}\ \pi_{+,\nu}(DF) + \mathrm{trace}\ \pi_{-,\nu}(DF)\}d\mu(\nu)$$

for all Schwartz class functions $F\colon P \to \mathbb{C}$.

§17. Fourier Inversion Inside the Group E_8

Finally, we apply the Fourier Inversion procedure to parabolic subgroups $P = MAN$ in a group G of type

E_8: 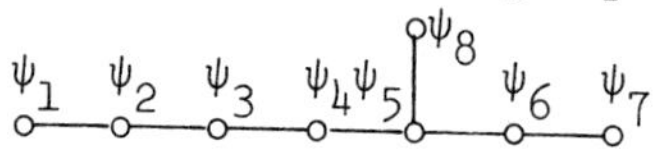

. The parabolics in question derive from $P_{\{\psi_1\}}$ and $P_{\{\psi_7\}}$. Each has nonabelian nilradical, so (6.6.5) and §7.5 give us

17.1.1. Theorem. *The following are the cases in which* $P = MAN$ *is a parabolic subgroup in a simple group* G *of type* E_8, N *has square integrable representations, and (this is automatic)* P *satisfies* (9.1).

1. $G = E_{8,D_8}$ *and* P *is isomorphic either to*

(17.1.2) $$\{\mathbb{R} + \mathbb{R}^{56}\}\cdot\{E_{7,A_7} \times \mathbb{R}^+\}$$

or to

(17.1.3) $$\{\mathbb{R}^{7,7} + \mathbb{R}^{64}\}\cdot\{Spin(7,7) \times \mathbb{R}^+\} .$$

2. $G = (E_8)_{\mathbb{C}}$ *and* P *is isomorphic either to*

(17.1.4) $$\{\mathbb{C} + \mathbb{C}^{56}\}\cdot\{(E_7)_{\mathbb{C}} \times \mathbb{C}^*\}$$

or to

(17.1.5) $$\{\mathbb{C}^{14} + \mathbb{C}^{64}\}\cdot\{Spin(14;\mathbb{C}) \times \mathbb{C}^*\} .$$

3. $G = E_{8,E_7A_1}$ <u>and</u> P <u>is isomorphic either to</u>

$$\{\mathbb{R} + \mathbb{R}^{56}\}\cdot\{E_{7,E_6T_1} \times \mathbb{R}^{+}\} \tag{17.1.6}$$

<u>or to</u>

$$\{\mathbb{R}^{3,11} + \mathbb{R}^{64}\}\cdot\{Spin(3,11) \times \mathbb{R}^{+}\}\ . \tag{17.1.7}$$

We run through the six cases.

<u>17.2</u>. $G = E_{8,D_8}$ <u>and</u> P <u>is given by</u> (17.1.2).

Here $Z = \mathbb{R} = \mathfrak{z}$, $\mathfrak{z}^* \cong \mathbb{R}$ under $\lambda(z) = \lambda z$, and $M = E_{7,A_7}$ acts trivially on $\mathfrak{z}^*$. We use the invariant polynomial

$$\psi(\lambda) = \lambda\ , \qquad \text{degree } d = 1. \tag{17.2.1}$$

The data k, ℓ, q of (9.3a) are 1, 56, 29. So we have a choice of invertible self adjoint operator D on $L^2(P)$,

$$D = (\partial/\partial z)^{29}\ , \quad \text{differential but not positive,} \tag{17.2.2a}$$

$$D = |\partial/\partial z|^{29}\ , \quad \text{positive but not differential.} \tag{17.2.2b}$$

The "unit sphere" in $\mathfrak{z}^*$ is $S = \{\pm 1\}$, two M-fixed points. Let $[\pi_\pm] \in \hat{N}$ denote the corresponding unitary representation classes. As $\pi_1(M) = \mathbb{Z}_4$, finite, they extend to classes $[\widetilde{\pi}_\pm] \in (NM)^\wedge$. So the "generic" representations of P are the

$$\pi_{\pm,\nu} = \mathrm{Ind}_{NM\uparrow P}(\widetilde{\pi}_\pm \otimes \nu), \qquad [\nu] \in \hat{M}\ . \tag{17.2.3}$$

That gives the Fourier Inversion formula

(17.2.4) $$F(1_P) = \int_{\widehat{E_{7,A_7}}} \{\text{trace}\ \pi_{+,\nu}(DF) + \text{trace}\ \pi_{-,\nu}(DF)\} d\mu(\nu)$$

for Schwartz class functions $F: P \to \mathbb{C}$.

<u>17.3</u>. $G = E_{8,D_8}$ <u>and</u> P <u>is given by</u> (17.1.3).

Here $Z = \mathbb{R}^{7,7} = \mathfrak{z}$, and $\mathfrak{z}^* \cong \mathbb{R}^{7,7}$ under $\lambda(z) = \langle \lambda, z \rangle$. $M = Spin(7,7)$ acts on $\mathfrak{z}^*$ by the vector representation

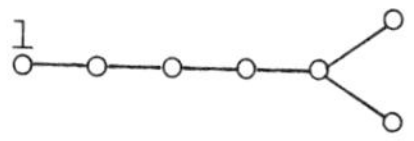
, which leaves invariant the polynomial

(17.3.1) $$\psi(\lambda) = \langle \lambda, \lambda \rangle, \qquad \text{degree } d = 2.$$

The data of (9.3a) are 14, 64, 46. So our invertible self adjoint operator on $L^2(P)$ is either of

(17.3.2a) $$D = \square^{23}, \qquad \text{differential but not postive}$$

(17.3.2b) $$D = |\square|^{23}, \qquad \text{positive but not differential}$$

where $\square$ is the Laplacian of $\mathbb{R}^{7,7}$.

The "unit sphere" $S = \{\lambda \in \mathfrak{z}^* : \langle \lambda, \lambda \rangle = \pm 1\}$ falls into two M-orbits,

(17.3.3) $$Ad^*(M) \cdot \lambda_{\pm} = \{\lambda \in \mathfrak{z}^* : \langle \lambda, \lambda \rangle = \pm 1\} \cong Spin(7,7)/Spin(7,6).$$

Let $[\pi_{\pm}] \in \hat{N}$ correspond to $\lambda_{\pm}$. Its M-stabilizer $M_{\pm} \cong Spin(7,6)$ has $\pi_1(M_{\pm})$ finite, so $[\pi_{\pm}]$ extends to class $[\widetilde{\pi_{\pm}}] \in (NM_{\pm})^{\wedge}$. Now P has "generic" representations

(17.3.4) $$\pi_{\pm,\nu} = \mathrm{Ind}_{NM_{\pm}\uparrow P}(\widetilde{\pi_{\pm}} \otimes \nu), \quad [\nu] \in \widehat{M_{\pm}} .$$

This leads to the Fourier Inversion formula

(17.3.5) $$F(1_P) = \int_{\mathrm{Spin}(7,6)^{\wedge}} \{\mathrm{trace}\ \pi_{+,\nu}(DF) + \mathrm{trace}\ \pi_{-,\nu}(DF)\} d\mu(\nu)$$

for Schwartz functions F on P which, in case of the choice (17.3.2b) for D, satisfy the additional conditions (9.5) on partial Fourier transform.

17.4. $G = (E_8)_{\mathbb{C}}$ and P is given by (17.1.4).

Here $Z = \mathbb{C} = \mathfrak{z}$, $\mathfrak{z}^* \cong \mathbb{C}$ under $\lambda(z) = \mathrm{Re}(\lambda z)$ and $M = (E_7)_{\mathbb{C}} \times \mathbb{C}'$ acts on $\mathfrak{z}^*$ by $(g,t)\colon \lambda \mapsto t^{-2}\cdot\lambda$. Our M-invariant $\mathbb{R}$-polynomial is

(17.4.1) $$\psi(\lambda) = |\lambda|^2, \qquad \text{degree } d = 2.$$

The data (9.3a) are 2, 112, 58. So the invertible positive self adjoint operator on $L^2(P)$ is

(17.4.2) $$D = \Delta^{29}, \qquad \text{differential,}$$

where Δ is the Laplacian of $Z \cong \mathbb{R}^2$.

The "unit sphere" in $\mathfrak{z}^*$ is $S = \{\lambda\colon |\lambda| = 1\}$, and the $\mathbb{C}'$ factor of M is transitive on it,

(17.4.3) $$S = \mathrm{Ad}^*(M)\cdot\lambda_1 \cong M/(E_7)_{\mathbb{C}} \times \{\pm 1\} .$$

Let $[\pi] \in \hat{N}$ correspond to λ_1. Here in the complex case extension is no problem: we get $[\tilde{\pi}] \in (NM_1)^\wedge$ and the "generic" representations of P are the

$$(17.4.4) \qquad \pi_\nu = \mathrm{Ind}_{NM_1 \uparrow P}(\tilde{\pi} \otimes \nu), \quad [\nu] \in \widehat{M_1}.$$

This leads to the Fourier Inversion formula

$$(17.4.5) \qquad F(1_P) = \int_{\{(E_7)_{\mathbf{C}} \times \{\pm 1\}\}^\wedge} \mathrm{trace}\ \pi_\nu(\Delta^{29} F)\, d\mu(\nu)$$

for Schwartz class functions F on P.

<u>17.5</u>. $G = (E_8)_{\mathbf{C}}$ <u>and</u> P <u>is given by</u> (17.1.5).

$Z = \mathbf{C}^{14} = \mathfrak{z}$ with $O(14;\mathbf{C})$-invariant bilinear form $\langle\ ,\ \rangle$, and $\mathfrak{z}^* \cong \mathbf{C}^{14}$ under $\lambda(z) = \langle \lambda, z \rangle$. $M = \mathrm{Spin}(14;\mathbf{C}) \times \mathbf{C}'$ acts on $\mathfrak{z}^*$ by $(g,t)\colon \lambda \mapsto t^{-2}\cdot\nu(g)\lambda$ where ν: 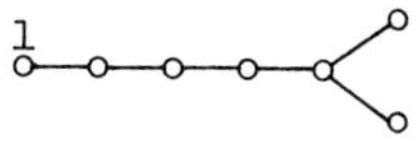 is the vector representation. The M-invariant $\mathbb{R}$-polynomial is

$$(17.5.1) \qquad \psi(\lambda) = |\langle \lambda, \lambda \rangle|^2, \quad \text{degree } d = 4.$$

The (9.3a) data are 28, 128 and 92, giving us the invertible positive self-adjoint operator on $L^2(P)$,

$$(17.5.2) \qquad D = \Psi^{23} = \Box^{46}, \qquad \text{differential},$$

where $\Box$ is the Laplacian of $\mathbf{C}^{14}$ as $\mathbb{R}^{14,14}$.

The "unit sphere" $S = \{\lambda \in \mathfrak{z}^*: |\langle\lambda,\lambda\rangle| = 1\}$ is a single M-orbit: fix λ_1 with $\langle\lambda_1,\lambda_1\rangle = 1$, some $g \in Spin(14;\mathbb{C})$ carries any given $\lambda \in S$ to a multiple $e^{2i\theta}\lambda_1$, then $t = e^{i\theta} \in \mathbb{C}'$ carries it on to λ_1. So

$$(17.5.3) \qquad S = Ad^*(M)\cdot\lambda_1 \cong M/Spin(13;\mathbb{C}) \times \{\pm 1\} .$$

The class $[\pi] \in \hat{N}$ corresponding to λ_1 extends immediately to $[\tilde{\pi}] \in (NM_1)^{\wedge}$, so the "generic" representations of P are the

$$(17.5.4) \qquad \pi_\nu = Ind_{NM_1\uparrow P}(\tilde{\pi} \otimes \nu), \quad [\nu] \in \widehat{M_1}.$$

That leads to the Fourier Inversion formula

$$(17.5.5) \qquad F(1_P) = \int_{\{Spin(13;\mathbb{C})\times\{\pm 1\}\}^{\wedge}} trace\ \pi_\nu(DF)d\mu(\nu)$$

for functions $F: P \to \mathbb{C}$ of Schwartz class.

<u>17.6</u>. $G = E_{8,E_7A_1}$ <u>and</u> P <u>is given by</u> (17.1.6).

Here $Z = \mathbb{R} = \mathfrak{z}$, $\mathfrak{z}^* \cong \mathbb{R}$ under $\lambda(z) = \lambda z$, and $M = E_{7,E_6T_1}$ acts trivially on $\mathfrak{z}^*$. The M-invariant polynomial is

$$(17.6.1) \qquad \psi(\lambda) = \lambda , \qquad \text{degree } d = 1$$

and the invertible self adjoint operator on P can be taken to be either of

(17.6.2a) $D = (\partial/\partial z)^{29}$, differential but not positive,

(17.6.2b) $D = |\partial/\partial z|^{29}$, positive but not differential.

The "unit sphere" in $\mathfrak{z}^*$ is $S = \{\pm 1\}$, two M-fixed points. Let $[\pi_\pm] \in \hat{N}$ denote the corresponding unitary representation classes.

<u>17.6.3. Lemma</u>. <u>The</u> $[\pi_\pm] \in \hat{N}$ <u>extend to classes</u> $[\tilde{\pi}_\pm] \in (NM)^\wedge$.

<u>Proof</u>. Let τ denote the 56-dimensional representation

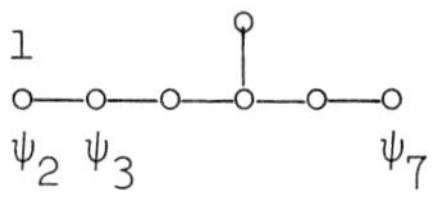 of E_7. We need the fact that

$\tau|_{E_6}$ = 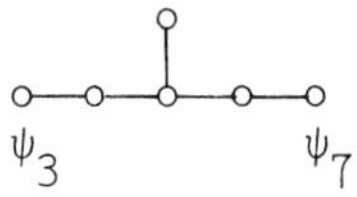 $\oplus$ 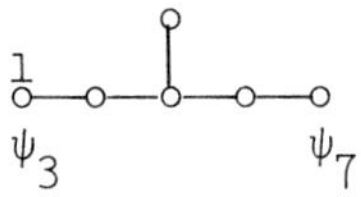 $\oplus$ 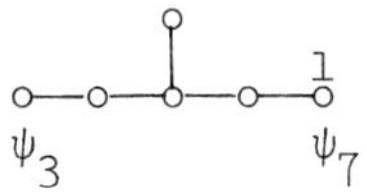 $\oplus$ 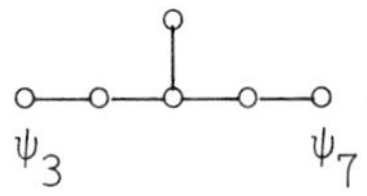,

sum of two trivial 1-dimensional representations and the two 27-dimensional representations of E_6. For that, let ξ denote the highest weight of τ. Then $\tau(E_6)$ is trivial on the highest weight space, $\xi - \psi_2$ is the highest weight for a subrepresentation of $\tau|_{E_6}$, and that subrepresentation is 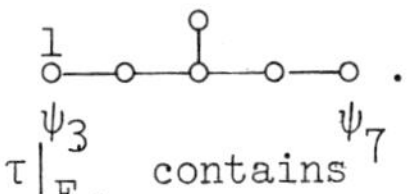. As τ is self-contragredient, so is $\tau|_{E_6}$, so now $\tau|_{E_6}$ contains 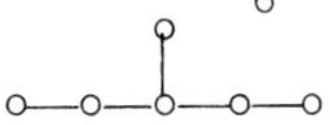 $\oplus$ 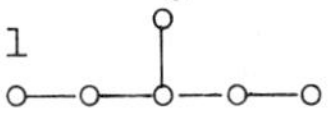 $\oplus$ 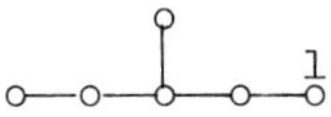. That leaves 1 dimension, which necessarily is a second 1-dimensional trivial representation of E_6.

Let $\overset{n}{x}$ denote the representation $e^{i\theta} \mapsto e^{in\theta}$ of the circle group T_1. We now check that

$\tau|_{E_6T_1} = \{$ 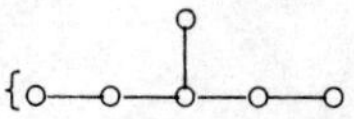 $\otimes\, x^3\} \oplus \{$ 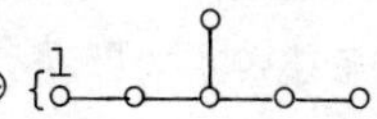 $\otimes\, x\} \oplus$

$\{$ 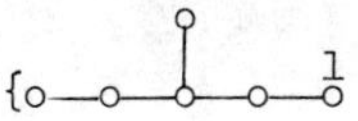 $\otimes\, x^{-1}\} \oplus \{$ 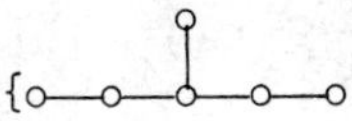 $\otimes\, x^{-3}\}$.

For that, use the E_7 tables at the back of Bourbaki [1] to verify $\|\xi\|^2 = 3\langle \xi, \xi-\psi_2\rangle$ and combine this with the decomposition of $\tau|_{E_6}$.

Now view the action of M on N/Z as a representation $f\colon E_{7,E_6T_1} \to Sp(28;\mathbb{R})$. On maximal compact subgroups it is $f\colon E_6T_1 \to U(28)$. There it is "half" of $\tau|_{E_6T_1}$, hence given on the T_1 factor either by

$$(17.6.4a) \qquad f(e^{i\theta}) = \begin{pmatrix} e^{3i\theta} & 0 \\ 0 & e^{i\theta}I_{27} \end{pmatrix} = e^{15i\theta/14}\begin{pmatrix} e^{27i\theta/14} & 0 \\ 0 & e^{-i\theta/14}I_{27} \end{pmatrix}$$

or by

$$(17.6.4b) \qquad f(e^{i\theta}) = \begin{pmatrix} e^{3i\theta} & 0 \\ 0 & e^{-i\theta}I_{27} \end{pmatrix} = e^{-6i\theta/7}\begin{pmatrix} e^{27i\theta/7} & 0 \\ 0 & e^{-i\theta/7}I_{27} \end{pmatrix}.$$

In case (17.6.4a) we have, in Proposition 13.2.6, $n = 28$, $m = 14$, $h = 3$ and $q = 15$, so $r = |nq/mh| = 10$. In case (17.6.4b) we have $n = 28$, $m = 7$, $h = 3$ and $q = -6$, so $r = |nq/mh| = 8$. Thus r is even in either case, and the assertion of the Lemma follows from Proposition 13.2.2.

q.e.d.

Now the "generic" representations of P are the

(17.6.5) $$\pi_{\pm,\nu} = \mathrm{Ind}_{NM\uparrow P}(\widetilde{\pi_\pm} \otimes \nu), \quad [\nu] \in \hat{M} .$$

That leads to the Fourier Inversion formula

(17.6.6) $$F(1_P) = \int_{\{E_7,E_6T_1\}^\wedge} \{\mathrm{trace}\ \pi_{+,\nu}(DF) + \mathrm{trace}\ \pi_{-,\nu}(DF)\} d\mu(\nu)$$

for Schwartz class functions $F: P \to \mathbb{C}$.

<u>17.7</u>. $G = E_{8,E_7A_1}$ <u>and</u> P <u>is given by</u> (17.1.7).

Here $Z = \mathbb{R}^{3,11} = \mathfrak{z}$, $\mathfrak{z}^* \cong \mathbb{R}^{3,11}$ under $\lambda(z) = \langle \lambda, z \rangle$, and $M = \mathrm{Spin}(3,11)$ acts on $\mathfrak{z}^*$ by the vector representation 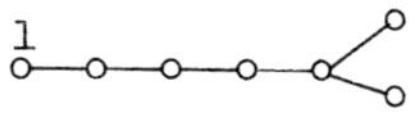.
The invariant polynomial is

(17.7.1) $$\psi(\lambda) = \langle \lambda,\lambda \rangle , \qquad \text{degree } d = 2.$$

In (9.3a), k, ℓ and q are 14, 64 and 46, so we have the choice of

(17.7.2a) $$D = \square^{23}, \quad \text{differential but not positive ,}$$

(17.7.2b) $$D = |\square|^{23}, \quad \text{positive but not differential ,}$$

where $\square$ is the Laplacian of $\mathbb{R}^{3,11}$.

The "unit sphere" $S = \{\lambda \in \mathfrak{z}^* : \langle \lambda,\lambda \rangle = \pm 1\}$ decomposes as union of two M-orbits

$$(17.7.3) \quad \begin{cases} Ad^*(M)\cdot\lambda_+ = \{\lambda \in \mathfrak{z}^* : \langle\lambda,\lambda\rangle = 1\} \cong Spin(3,11)/Spin(2,11), \\ Ad^*(M)\cdot\lambda_- = \{\lambda \in \mathfrak{z}^* : \langle\lambda,\lambda\rangle = -1\} \cong Spin(3,11)/Spin(3,10). \end{cases}$$

Let $[\pi_\pm] \in \hat{N}$ correspond to $\lambda_\pm \in \mathfrak{z}^*$.

<u>17.7.4. Lemma</u>. <u>Each</u> $[\pi_\pm] \in \hat{N}$ <u>extends to a class</u> $[\widetilde{\pi_\pm}] \in (NM_\pm)^\wedge$.

<u>Proof</u>. Since $\pi_1(Spin(3,10)) = \mathbb{Z}_2$, finite, $[\pi_-]$ extends to a class $[\tilde{\pi}_-] \in (NM_-)^\wedge$.

Let σ denote the (half spin) representation o—o—o—o—o< (upper branch node labelled 1) of M on N/Z. Then σ restricts on $M_+ = Spin(2,11)$ to the spin representation o—o—o—o—o⇒o (last node labelled 1), and its further restriction to the maximal compact subgroup $Spin(2)\cdot Spin(11)$ is

$\{\frac{1}{x} \otimes$ o—o—o—o⇒o $^1\} \oplus \{\bar{x}^1 \otimes$ o—o—o—o⇒o (last node labelled 1)$\}$. Parameterize the 1-parameter group $Spin(2)$ as $\{r(\theta) : 0 \leqslant \theta \leqslant 2\pi\}$ and view $\sigma|_{M_+}$ as a map $f: Spin(2,11) \to Sp(32;\mathbb{R})$. On maximal compact subgroups, $f: Spin(2)\cdot Spin(11) \to U(32)$. As $f \oplus \bar{f} = \sigma|_{Spin(2)\cdot Spin(11)}$, now $f(r(\theta)) = e^{i\theta} I_{32}$. Thus, in Proposition 13.2.6 we have $n = 32$, $m = 1$, $h = 2$ and $q = 1$, so $r = |nq/mh| = 16$, even. Now Proposition 13.2.2 guarantees the existence of $[\tilde{\pi}_+]$.

<u>q.e.d.</u>

Now the "generic" representations of P are the

$$(17.7.5) \qquad \pi_{\pm,\nu} = Ind_{NM_\pm \uparrow P}(\widetilde{\pi_\pm} \otimes \nu), \quad [\nu] \in \widehat{M_\pm}.$$

That leads to the Fourier Inversion formula

$$(17.7.6)\quad \begin{cases} F(1_P) = \displaystyle\int_{\mathrm{Spin}(2,11)^\wedge} \mathrm{trace}\ \pi_{+,\nu}(DF)\,d\mu_+(\nu) \\ \qquad + \displaystyle\int_{\mathrm{Spin}(3,10)^\wedge} \mathrm{trace}\ \pi_{-,\nu}(DF)\,d\mu_-(\nu) \end{cases}$$

for Schwartz class functions $F: P \to \mathbb{C}$ which, in case of the choice (17.7.2b), satisfy the additional conditions (9.5) on partial Fourier transform.

References

[1] N. Bourbaki, "Groupes et Algèbres de Lie, Chapitres 4, 5 et 6," Hermann, Paris, 1968.

[2] H. Braun u. M. Koecher, "Jordan Algebren," Springer-Verlag, Berlin, 1966.

[3] J. Charbonnel, La formule de Plancherel pour un groupe de Lie résoluble connexe, thèse, Université de Paris, 1975.

[4] D. Drucker, "Exceptional Lie algebras and the Structure of Hermitian Symmetric Spaces," Memoirs Amer. Math. Soc. **208**, 1978.

[5] M. Duflo and C. C. Moore, On the regular representation on a non-unimodular locally compact group, J. Funct. Anal. **21** (1976), 209-243.

[6] N. Jacobson, "Structure and Representations of Jordan Algebras," Amer. Math. Soc. Colloq. Publ., Providence, 1969.

[7] N. Jacobson, "Lectures on Quadratic Jordan Algebras," Tata Institute, Bombay, 1969.

[8] F. W. Keene, Square integrable representations and a Plancherel theorem for parabolic groups, Trans. Amer. Math. Soc. **243** (1978), 61-73.

[9] F. W. Keene, R. L. Lipsman and J. A. Wolf, The Plancherel formula for parabolic subgroups, Israel Math. J. **28** (1977), 68-90.

[10] A. Kleppner and R. L. Lipsman, The Plancherel formula for group extensions, Ann. Sci. É.N.S. **5** (1972), 459-516.

[11] A. Kleppner and R. L. Lipsman, The Plancherel formula for group extensions, II, Ann. Sci. É.N.S. **6** (1973), 103-132.

[12] A. Kohari, Harmonic analysis on the group of linear transformations of the straight line, Japan Academy--Proceedings **37** (1961), 250-254.

[13] A. Korányi and J. A. Wolf, Realization of hermitian symmetric spaces as generalized half-planes, Annals of Math. **81** (1965), 265-288.

[14] R. L. Lipsman and J. A. Wolf, The Plancherel formula for parabolic subgroups of the classical groups, Journal d'Analyse Math. **34** (1978), 120-161.

[15] O. Loos, "Jordan Pairs," Lecture Notes in Math. **460**, Springer-Verlag, Berlin, 1975.

[16] K. McCrimmon, Jordan algebras and their applications, Bull. Amer. Math. Soc. **84** (1978), 612-627.

[17] C. C. Moore, Representations of solvable and nilpotent groups and harmonic analysis on nil- and solvmanifolds, Proc. Symp. Pure Math. 24 (1973), 3-44.

[18] C. C. Moore and J. A. Wolf, Square integrable representations of nilpotent groups, Trans. Amer. Math. Soc. 185 (1973), 445-462.

[19] L. Pukánszky, Unitary representations of solvable Lie groups, Ann. Sci. É.N.S. 4 (1971), 464-608.

[20] W. Schmid, Die Randwerte holomorpher Funktionen auf hermitesch symmetrischen Räumen, Invent. Math. 9 (1969/70), 61-80.

[21] S. Sternberg and J. A. Wolf, Hermitian Lie algebras and metaplectic representations, I, Trans. Amer. Math. Soc. 238 (1978), 1-43.

[22] N. Tatsuuma, Plancherel formula for non-unimodular locally compact groups, J. Math. Kyoto Univ. 12 (1972), 179-261.

[23] J. A. Wolf, On the classification of hermitian symmetric spaces, J. Math. and Mech. (later called Indiana Univ. Math. J.) 13 (1964), 489-496.

[24] J. A. Wolf and A. Korányi, Generalized Cayley transformations of bounded symmetric domains, Amer. J. Math. 87 (1965), 899-939.

[25] J. A. Wolf, Fine structure of hermitian symmetric spaces, in "Symmetric spaces," ed. Boothby and Weiss, Dekker, New York, 1972; 271-357.

[26] J. A. Wolf, "Unitary Representations on Partially Holomorphic Cohomology Spaces," Memoirs Amer. Math. Soc. 138, 1974.

[27] J. A. Wolf, Representations of certain semidirect product groups, J. Funct. Anal. 19 (1975), 339-372.

[28] J. A. Wolf, "Unitary Representations of Maximal Parabolic Subgroups of the Classical Groups," Memoirs Amer. Math. Soc. 180, 1976.

[29] J. A. Wolf, "Spaces of Constant Curvature (4th edition)", Publish or Perish (P.O. Box 7108, Berkeley CA 94707), 1977.

[30] J. A. Wolf, Representations associated to minimal co-adjoint orbits, in "Differential Geometrical Methods in Mathematical Physics II (Proceedings, Bonn 1977)," Springer Lecture Notes in Mathematics 676 (1978), 329-349.

Department of Mathematics
University of California
Berkeley, California 94720